A Body of Water

An Accidental Education in Evolution

M. T. Harber

Mainsail Breeze Edition, 2019
Copyright © 2019
All Rights Reserved

Mainsail Breeze
ISBN 978-0-9835898-9-1
WWW.MAINSAILBREEZE.COM

Other Books By M.T.Harber

The Sweet Taste of the Bilge

The Eye of the Abyss

Hemingway's Storm

The Reverse of Babylon

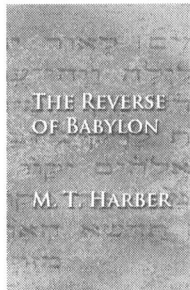

Soylent Blue

DEDICATION

This book goes out to all who have some inexplicable draw to the water. For those who do not have this feeling, you will never know. It can't be explained, it can't be reasoned away. I've shared the water dreams with many people, including my wife, Jeanette. I have a picture of her looking an an angel fish on one of our first dives. They aren't there in such numbers anymore. Much of the ocean glory has disappeared. But the magic remains. I've tried to get the science right. Where I have failed, I only blame myself.

I'd like to thank all of those who taught me how too SCUBA; Jim Aurich, Judy and Terry Brady, Steve Thompson, Claude Smith, Steve Dye, Jim Bates, Woody Alpern, Bev, Wyatt and Chelsea Foster, John Player, Brian Acker, Dan Weekly, Winston von Rensburg, There will be others I've forgotten - dive buddies that I've shared a moment of wetness and a drink or three.

For those of you who have yet to dip their toe in the wonderful blue any deeper than you can hold your breath, you are still invited to walk - or swim - with me along this journey. I hope you seek me out to share your stories and provide insight in this adventure that I stumbled upon.

INTRODUCTION

You didn't come into this world. You came out of it, lie a wave from the ocean. You are not a stranger here.
        — Alan Watts

It was the perfect day - well, for me it at least. I was the youngest of seven children, all of twelve years old, and now I had my father all to myself. My father worked in a mine, leaving before dawn and coming home late, so this moment was precious time indeed. The family was on vacation at Rehobeth

Beach, Delaware. We'd been there about three days when the weather turned. The sun worshippers had all gone inside. A front was moving in with the afternoon heat. With the changing of the tide and a storm out to sea, the lifeguards had shut down the beach.

Three hours ago it was a completely different scene. The once blue sky was now a steel grey overcast. The mares' tails travelled overland and darker cumulus clouds were gathering out to sea. My father and I had been body surfing the waves all day. Now the waves were coming with greater rapidity. The height of the waves had doubled. I could feel the undertow pulling us down the beach.

My mother had argued briefly with my father and taken my sister inside. Dad and I were soaring along the crest of wave after wave. My chest was rubbed raw as I crashed through the foamy remnants of a massive curl. The boogie board was my only hope for reaching the surface and getting a breath before I was rolled in a sandy mix. Dad was having as much fun as me. My adrenaline surged as we dove under wave after crashing wave to find the sweet spot where I couldn't feel the bottom as the ocean birthed waves with violent regularity.

"Take this one!" my dad yelled and I pumped my spindly legs furiously to get ahead of the mounting crest. I felt the break and moved in that direction, trying to stay in the thread of the power that threatened to overtake me.

My dad wasn't far behind slamming down from a wave that showed no mercy.

I didn't feel fear. Fear destroys the mind and shuts down the senses. I was energized and in-tune with both the ocean and my father. It was one of those unscripted moments between father and son that ignites a deeper understanding and respect of the entire universe and our shared place in it.

Eventually the lightning and dangerous rip currents forced us out of the water. We were both bruised, bleeding and smiling from ear-to-ear. The rain started in thick sheets and by the time we returned to the rental home the sky was dumping buckets from some unseen boxcar overhead. I was shivering, winded, blue-lipped and in need of dry clothes as I stepped into the rental cabin along the beach. But, I knew that I'd experienced a perfect day at the beach.

The bonding moments between me and my father usually centered around some kind of aquatic event. Whether it was my father throwing me into a pool as a kindergartener or heading out before sunrise on a head-boat to go fishing my life seemed destined to find my solace out in the big blue.

For years I would talk of days where I had "Island head" - always thinking about the beach, the water and a way of life far different from the one I was currently living. As a teen, I spent many days with friends seeing who could hold their breath underwater the longest. I learned to blow bubbles out of my eyelids. I blew rings with precision and watched them grow as they rose to the surface. I wasn't even aware such a thing as free diving existed until I chanced upon the movie *The Big Blue*.

This mythical interpretation of the free diving brotherhood between Jacques Mayol and Mario Enzo solidified my calling to the water. It was more than a psychological compulsion. There seemed a physical pull to the water. When I was in the water, things changed. I felt radically different. There was a physiological component that went beyond the psychological advantages found in such books as *Blue Mind, Neutral Buoyancy,* and *One Breath*.

It was in the summer of 2015 I found myself reveling in the post-dive high as I headed north on the "PADI Wagon" - a 1960's tour bus - after four days of SCUBA diving in the Florida Keys. There was a distinct olfactory "old fish" funk as twenty diver's gear hastily and improperly rinsed mixed with the rampant burning of oil and diesel all cooked by the unbridled sun of a hot Florida day and a malfunctioning air-conditioner. Such a smell creates a habitat for unseen growth you can't dismiss with the ten hanging green pine tree fresheners located along the length of the vehicle. Many of the divers carried onboard their own distinct smell – left over from foregoing the military rinse after the dive in favor of maximizing their time with the tribe for the post-dive cookout before heading on the long trek North to Atlanta Georgia. The majority of the divers on the PADI Wagon ignored this nasal assault - and for good reason.

This was a distinctly mellow and quite different crowd that left Atlanta four days prior. Departing Atlanta on a Thursday evening the newly initiated divers heading down to complete their Open Water certification were quick with nervous banter. Those who'd long since completed this initiation of check-out dives told tales of memorable locations where some odd inhabitant of the deep rose to greet them, or they relayed some challenge they conquered during a technical dive. Once the beer cans were opened and the rum and cokes made their rounds, the volume of banter increased and the night trip was filled with raucous laughter.

Now, with seven dives completed and everyone fatigued from the long weekend it was a quiet but tight-knit group of like-minded souls who shared a tribal SCUBA experience. The group heading back was exhausted. The younger ones found a bunk or seat and curled up in a sweaty ball. The driver was listening to an e-book and the general haze of exhaustion pervaded the bus like the tropical humidity before a storm.

I was both tired and elated but nowhere near ready to crane my neck back in search of elusive rest. I was still trying to sustain my high from the morning's dive. Sea turtles nibbled along green outcrops while a small nurse shark darted overhead. The fan coral swayed back and forth to the rhythm of the surge like a conductor of an orchestra as the wrasse and sergeant majors followed in unison undulation. Even though it was a working vacation, there were moments when I could pause and hover weightless over a coral forest while hundreds of fish, reptiles and mammals performed the delicate ballet of life and death that is the aquatic ecosystem of Key Largo.

Those not initiated to the world of SCUBA diving often misconstrue diving as a high energy extreme sport. Quite the opposite, the goal of a good dive is relaxation, air conservation and unobtrusive observation. Others on the bus and the SCUBA tribes with whom I align follow the same credo that diving isn't so much a sport as it is therapy. There is something cathartic living and moving in three dimensions in an atmosphere that cradles you like a mother's womb.

Now I was back on land. I'd loaded all of the equipment onto the bus. I was the divemaster – aka grunt – who had to make sure the equipment and luggage was all onboard. I no longer resembled that wave chasing child who rode the crashing waves, I was now bald and counted among the ranks of "old men."

And, like an old man who hauled twenty sets of SCUBA gear, my back was killing me. On the "PADI Wagon" there was no place that a man standing six foot four inches could properly stretch to alleviate the pain and soreness I was feeling. In that moment of post-SCUBA satori and physical distress I entered some sort of altered state of consciousness. Maybe it was the smell – like entering the SCUBA version of a hookah bar. In the water my feet felt fine, my back didn't hurt and things made sense. The "real" world, on land, I was out of my element, an air-breathing fish-out-of-water.

During a dive in the Gulf of Mexico in Panama City a few months prior, a fellow diver was suffering from incredible back pain. There was no way this man could lift his equipment on his back. He was barely able to lift a water bottle to his lips. He was bent over as if he was auditioning for the role of Ichabod Crane. Was that the end of his day of diving? No! On the contrary, getting this man in the water was the balm of Gilead that he most needed. It was the healing pool, the magic juice, and the miracle he needed. We made his equipment positively buoyant and tossed it off the boat. We got in the water with him and helped him don his gear. He joined us as we submerged into the murky blue-green waters as the current threatened to pull him from the mooring line. Once at the bottom the weightless conditions stretched his compressed vertebrae. By the end of the dive he was well enough to pull himself out of the water and onto the boat with all of his gear. He was totally healed and quite renewed.

Somewhere along the palm-tree lined Highway One of the Florida Keys I began wondering if there was some evolutionary common-sense related to that healing that day. Was there an evolutionary path that took us along the shoreline and back to the aquatic realm? And, if we did take a quick detour like many other land mammals who returned to the sea, is there evidence of this occurrence? Were we the product of evolutionary indecision? If so, where would this evidence of our aquatic evolution reside? And, is my aching back part of the evidence? It seemed unlikely.

But it was not. In fact, the evidence is there – everywhere – in everyone.

Then I started thinking about all of my "earthly" quirks. Sore feet, aching back, baldness (yes I own that), walking upright, big heads, odd "human" toes, and a myriad of other traits that made a human distinct from other animals and more specifically, other primates. I didn't have Google on the SCUBA-bus to satisfy my curiosity. Perhaps that was for the best. Sometimes Google is like a Hot Pocket. It will take care of your hunger so you forget about that discomfort in your gut. But it also isn't the best thing for you. A proper meal takes time. So does deep rumination of our origins.

MORGAN'S MONKEYS

The trouble with specialists is that they tend to think in grooves. From time to time something happens to shake them out of that groove.
— Elaine Morgan

The most vivid recollection from my many trips to the Smithsonian Museum of Natural History was of a dimly lit room filled with human skulls. As a kid I felt a mix of fear and curiosity as my ancestors' bones stared back at me through hollow eyes. The most ancient, brown and decayed small skulls sat near the bottom of the display and each generation sat above their predecessor gazing back in an orderly progression with the most recent incarnation atop the stack. The display was the human discovery of some divine humanoid experiment leading its grand conclusion with the perfected "us" at the pinnacle of the glass case. The backdrop of humanity's progression was illustrated as a series of painted arrows marking the evolution across the globe spanning millions of years. There was a solemnity to this section of the museum. We were walking among the dead as the skulls stared defiantly, daring a child to peer into their remnants and divine some deeper meaning of those grandfathers of old.

Along the opposite wall were a series of dioramas. Each cove displayed a different milestone in human evolution. The artists took the scientific understanding of the evolutionary tree and rendered them in life-sized three dimensional forms. The early hominids were short. Mixed with them were animals they hunted and a terrain of scrub and thin trees scattered throughout. The savanna backdrops looked sparse with a small line of trees on a distant horizon. A volcano's smoke billowed beyond them. The small figure in the foreground crouched amidst a tangle of scrub, fearfully scanning the distance. The figure seemed more animal than human - barely recognizable in the human lineage. A little creek artfully dotted the corner of the display. Each subsequent rendering built upon the same theme. As a child, I accepted these representations as fact. This is they way it was. I never questioned the accuracy of these representations. They matched the pictures I'd seen in my science books at school. The savanna theory of evolution was the accepted standard upon which all of the evolutionary science was placed.

But in the 1970's and 80's a Welsh television writer named Elaine Morgan challenged the scientific community with a new ideas about our evolution. Her first book, *The Descent of Woman* (1972) challenged the male-dominated theories of evolution and elevated the role of the woman as merely "child-bearer" to an equal role in the evolutionary plan. The controversial work became a best seller - more for its role in feminist literature than a new treatise on evolution. It was a slap in the face to the earlier controversial book *The Naked Ape* by Desmond Morris. Morris articulated throughout the book that it was the male hunter that allowed the species as a whole to progress and evolve. What was the role of women?

"The females found themselves almost perpetually confined to the home base[1]."

That's why women have breasts, Morris contends. He articulated with a deep vocabulary, the nakedness of early hominids was the attraction and, therefore, the emperor's clothes that propelled him to the top of the food chain. It was the male who directed the course of evolution according to the book. It is an interesting read from the standpoint of comparative anatomy, but its assumptions have a distinct smell of 1960's male ego. Scientific research, even during Moriss's time was defusing the male-centric notion of the hunter as dominant in the family dynamic. His zoological expertise is focused to prove his point while ignoring animals like the bonobo, who have a highly sophisticated social system, demonstrate sophisticated tool use, and - by the way - exists as a matriarchal society. He should have done a little more reading on killer whales, elephants, and meerkats (while not even considering the insect world).

It was Morris and his followers who were early protagonists for Elaine Morgan's ideas. The battle for relevance continues though Morgan died a few years back. Morgan's lifetime argument against the Smithsonian-esque savanna theory dioramas were simple and to the point:

"I believe this is a mistake. The legend of the jungle heritage and the evolution of man as a hunting carnivore has taken root in man's mind as firmly as Genesis ever did… He has built a beautiful theoretical construction, with himself on the top of it, buttressed with a formidable array of scientifically authenticated facts. We cannot dispute the facts. We should not attempt to ignore the facts. What I think we can do is to suggest that the currently accepted interpretation of the facts is not the only possible one. "[2]

In *The Descent of Woman,* Morgan goes on to eviscerate many universally held beliefs, challenging tool use by hunters, sexual prowess and a host of other norms in the scientific community. Had it come from a male established scientist, it would have garnered praise. However the scientific community responded with mild disdain. It wasn't regarded as real science. The idea was cast onto the pile of new-age quack literature with the likes of *Chariots of The Gods* by Erich von Däniken and *Journey to Ixtlan: The Lessons of Don Juan* by Carlos Castaneda. This was a time of ancient astronauts and searching for the Loch Ness monster. New-Age spiritualists/scientists were proclaiming evidence for astral projection, UFO's, ghosts, the Bermuda Triangle, and pyramid power without a shred of true scientific rigor.

The age of pseudo-science was on the rise and science itself was becoming consumed by capitalists fabricating results for the corporate product. Eggs were touted as bad for you in favor of a synthetic egg substitute. The Dairy industry influenced the government to alter the "food guide pyramid" to a degree that the diet they suggested was actually bad for you.

This environment created an uphill battle in what was considered a new-age crossover subject. Morgan reevaluated the science currently available within her grasp and reinterpreted the established paleontological dogma upon which countless papers had been written.

"… As the heat and dryness spread out from the baking heart of Africa, it became reduced to a narrowing strip; the larger and fiercer arboreans drove her away…she was also hampered by a clinging infant; and she also was chased by a carnivore and found there was no tree she could run up to escape. However, in front of her there was a large sheet of water. With piercing squeals of terror she ran straight into the sea. The carnivore was a species of cat and didn't like wetting his feet; and moreover, though he had twice her body weight, she was accustomed like most tree-dwellers to adopting an upright posture, even though she used four legs for locomotion. She was thus able to go farther into the water than he could without drowning…. "[3]

Morgan's research led her to rethink the facts and postulate a new paradigm for human evolution. Taking what science provided, she assimilated as series of papers and tenets of evolution into a plausible new scenario. Ten years later she wrote, *The Aquatic Ape: A Theory of Human Evolution.*

"If we regard the ancestral primate as an aquatic ape, he ceases to be a mysterious zoological aberration evolving unique and inexplicable features of no use to himself and highly deleterious to his children. Put him among the aquatic mammals and he becomes a conformer"[4]

In fact, some scant references to this notion of early man as a creature living near water and accessing that habitat go back to 1912. The synthesis of ideas started with Max Westenhöfer, a German pathologist who pondered the notion in his treatment *Der Eigenweg des Menschen* (The Path Travelled by Man Alone.) The intent of the treatise was to present that which made man unique among animals from a physiological standpoint. This text was odd at best, meandering and postulating a number of far-fetched ideas.

It was, in its day, considered an honorable scientific treatise. But sometimes Westenhöfer went off the deep end. In particular, Westenhöfer focused on the shape of the human foot and the absence of hair (or fur). But later in the text Westenhöfer wrote about the possibility of Beowulf being history rather than fantasy. He meanders into a discussion about the possibility of man fighting dragons along the shore. His X-Files digression tainted an otherwise interesting piece of work that could've otherwise stood scientific scrutiny. His use of pathology and forensic research was marred by his dip into a Tolkein-like realm that had no place in scientific research.

Independently of Morgan and Westenhöfer, a marine biologist and SCUBA diver named Alister Hardy, who got knighted for his research on plankton, focused on an aquatic connection to hair loss, but as a member of scientific community in the '20s, he also included telepathy as part of his research on evolution. The guy was smart, but dabbling in the psychic realm tarnished any hope of taking his research seriously. Elaine Morgan read his less fringe-worthy articles about the aquatic connection and she was hooked.

Morgan hoped to dispel some of the aura of the supernatural mumbo jumbo underpinning this quiet theory and lay the theory of the "aquatic ape" in a sound scientific bedrock. Her book highlighted those elements of our "humanness" that separated us from both our primate cousins and our terrestrial kin in general.

Injecting common sense and scientific inquiry, Morgan provided alternatives to many universal beliefs of human evolution by playing compelling "what if" games. The entrenched scientists' ideas ran deep and they were quick to respond to Morgan's reinterpretation of the facts. Morgan's lack of credentials set her as a target for rebuttal and dismissal by those who made their living supporting theories they'd devised. Arguments rang similar to this one written by Henry Gee, an editor and contributor to *Nature* magazine:

"Hardly a month goes by without my receiving, at my desk at Nature, an exegesis on the reasons how or why human beings evolved to be this way or that. They are always nonsense, and for the same reason. They find some quirk of anatomy, extrapolate that into a grand scheme, and then cherry-pick attributes that seem to fit that scheme, ignoring any contrary evidence. Adherence to such schemes become matters of belief, not evidence. That's not science – that's creationism. An arcane and eldritch item of fashionable nonsense in the shuddering slush-pile of human evolutionary rubbish is the "aquatic ape theory"[5]

Time after time I have read books, reports, and scientific journals dismissing Morgan's theory. They seem to have found a singular source - Like Nina G. Jablonski's article in *Scientific American: The Naked Truth - Recent findings lay bare the origins of human hairlessness—and hint that naked skin was a key factor in the emergence of other human traits.* In the article she lays out a special blurb to dispel Morgan's theory. Her reasoning is as follows:

- Ocean mammals don't resemble Morgan's description
- The Aquatic Ape Theory is complicated

- There were crocodiles[6]

Little more reasoning is given other than a quick dismissal. Like any theory and scientific inquiry, it must be poked and prodded. It must be written and rewritten with the advent of new scientific discoveries. In fact, by the same token, the savanna theory struggles to stretch our human quirks into a working model. Looking at bipedalism, for example, the scientific consensus is that we became a primate who hurled objects at our predators. The next time you're faced with a predator bearing down on you, stop, pick up a rock, and throw it at them. Then you can let me know how sensible this theory sounds.

Yes, there were crocodiles. There were also land-based predators.

As for its complexity, we are talking about an alternate unifying theory using the same science and same set of facts. As scientific inquiry opens new paleontological evidence, it is the data that should drive the direction of the inquiry - no matter how far-fetched the proposal.

There are other theories contending for attention but each one makes a point to squelch the Aquatic Ape Theory. Morgan - never wavering - spent the rest of her life under the cross examination of scientists favoring to preserve their own idea.

More often than not, scientists used their credentials, rather than evidence to dispel Morgan. That is not to say that there weren't problems with the evidence supporting her theory. But that is true with any scientific journey. The argument is to take the hypothesis and either build it up or tear it down as new evidence comes to light.

Many misunderstood her theory, believing that Morgan's vision presented a primate turned manatee. Sometimes, it was simply a misunderstanding of Morgan's intent that led scientists to a quick dismissal. Some believed that Morgan was talking about the aquatic ape theory as a return to the ocean as if we were at one point mermaids and mermen swimming like fish. This was certainly *not* Morgan's theory. The mermaid mistake only fed the new-age movement and further distanced her ideas from scientific plausibility. It was as if

Westenhöfer's and Hardy's ghosts had possessed the science with their psychic folly. Morgan, however had the upper hand in one, very important, way - media.

In 1998, BBC and the Discovery Channel teamed to present her ideas in a well-crafted television documentary. This was followed by other television and radio programs. Other scientists didn't have the mainstream publicity, relegated to the handful of scientific journals mostly ignored by the general public. It is not to say that the shows were 100% accurate. But it put the scientific community on the defense. It also meant that every scholarly article had to have a caveat like that found in the Scientific American quote mentioned above, having to footnote a disclaimer for their evidence in light of those adhering to the plausibility of the Aquatic Ape Theory.

In some way the notoriety for Morgan's hypothesis worked in her favor. It provided continuing viability and voice. In other ways it worked against her. Instead of introducing an alternative that would be debated seriously, many in the scientific community dismissed the idea out of hand without fully understanding what she was proposing.

Scientists were busy celebrating "Lucy" the famous Australopithecus, found in Ethiopia. There were other fossil remains found from this era, but nothing so complete and well preserved. It showed the larger brain and stance this early figure possessed. Those supporting the savanna theory delved into the details to support their respective theories while dismissing Morgan.

However, what is little reported by those same scientists is the fact that Lucy was found, surrounded by crocodile eggs, turtle eggs, and crab claws, playing into and supporting Morgan's Theory.

Science has changed considerably since Morgan brought her ideas to the mainstream. Genetics has expanded new avenues to support or dispel Morgan's ideas. New remains and fossil evidence have forced us to rethink the human line as not so unique and straightforward as some textbooks would make you believe.

Some scientists, researchers, authors and enthusiasts have picked up where Elaine Morgan has left off. Marc Verhaegen, a general medical practitioner from Belgium wrote a number of articles supporting the Aquatic Ape Theory with new discoveries. Algis Kuliukas is another researcher who has contributed to the dialogue, presenting an interpretation similar to that found here.

Yet, like a tightly held secret society, the scientists holding fast to the savanna model regularly besieged anyone attempting to dissuade them from their firmly held beliefs. But, when looking at the sum of what our human evolution has brought forth, it falls far short of the mark. It is easy to target one or two features and build stories around them. Like physicists working on a a unified theory, Elaine Morgan was never afraid to connect the dots in new and challenging ways.

Science is an ever blooming flower, offering new revelations and rewriting the story. Will new insights exonerate Morgan and her predecessors or put to rest the hypothesis of the Aquatic ape? One should pause and simply contemplate the end result of human evolution: the human body. The scars of our past run deep in our very being and lead back to some interesting origins.

FOOT NOTES AND FALLEN ARCHES

Everywhere is within walking distance, if you have the time.
— Steven Wright

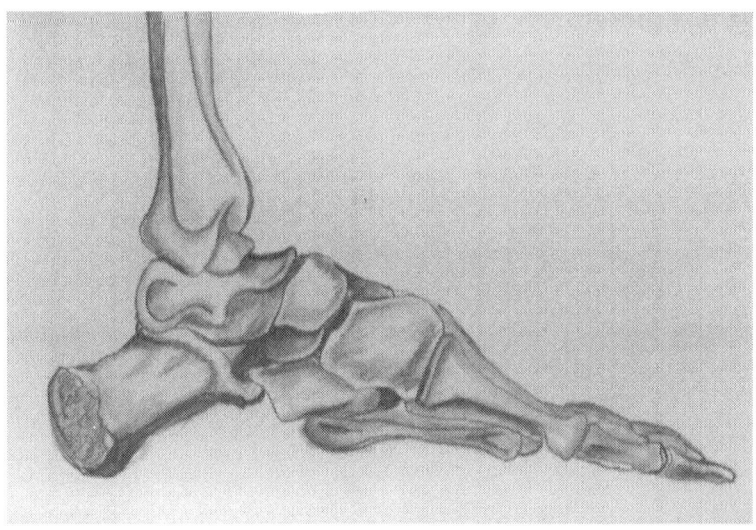

Ever stand on your toes to get something that was otherwise out of reach? You rise up for that extra inch that spells the difference between getting the sugar container from the top shelf or creating a sticky disaster. But did you know that standing on your toes may be evidence of a significant evolutionary break, providing something profound and overlooked in our hominid past?

Over the last few million years, our hands have maintained a distinct similarity to our primate brethren. With our hands we manipulate everything from grasping pencils to tying shoe laces. Photograph the palms of humans and orangutans and some may have difficulty telling them apart. However, the same people will have no trouble identifying human and orangutan feet. Whereas orangutan feet and hands are nearly identical, our feet are but a dim reference to our evolutionary cousins. Primates easily pick up sticks with their feet or jostle overhead along a dark canopy gracefully grabbing libs with all four "hands." We hide our pedal vestiges inside a pair of sneakers.

Our feet are architecturally different than any other animal. Highlighting this difference is our "big toe" or Hallux Major or "foot thumb" whatever you want to call that giant among dwarf toes on your feet. The big toe (I'm foregoing the scientific name, but feel free to insert whatever name or nickname you've given those jolly giants of the pedicure) is aligned with the rest of the digits and not opposed like the thumb on our hand.

While some human amputees have built up dexterity with the big toe, the majority of us rely on it solely for balance and propulsion. Those with "turf toe", bunions, or other maladies of the big toe know how painful walking becomes, not only because of the point of injury, but the pain of altering the entirety of the human bipedal mechanism to accommodate the injury. Primarily the arch of the foot (plantar facia), small muscles in the ankles, and adjustments in the knees take on a new painful role to adapt. Additionally, the ability to roll the foot during walking (starting with a heel strike and ending with the big toe "push off") becomes something that more resembles our bonobo clan's version of the two-legged stride.

We are a body in constant motion, even our circulatory system and breathing have an effect on our feet. We constantly adjust our stance and are well aware of the foot problems that occur when we try to maintain a static vertical position for too long. Hamstring cramps are not uncommon in the early spring as we pull our foot out of hard shoes (whose "sole role" is to disperse the requirements of the big toe's role in balance) and put on our flip-flops. Suddenly, the big toes is now responsible, not only for balance, but for pulling the flip-flop during walking. This is why doctors report an increase of foot and leg injuries in early spring.

If you are still wondering about the importance of the big toe in your balance, perform this experiment. Stand somewhere where you can grab a support (table, chair etc... you'll need it!). With your hands free, lift up a leg and close your eyes. While attempting to balance, feel how the big toe becomes the conductor of your balance. If you last thirty seconds, you should join a ballet company. Notice how the entire leg starts to play a role in this balancing process.

The essence of walking - that is bipedalism - isn't really the notion of two footed walking. In fact it is one footed walking in that the walking process requires us to be on one foot to propel us forward. Not only that, the final segment of the one-foot-on-the-ground activity means that there is a static moment where the big toe holds the weight of the body before the opposing foot reaches the ground.

Evolution of the foot

Paleoanthropologist Yohannes Haile-Selassie wasn't aware he'd discovered a new species while exploring Ethiopia during a 1975 expedition. But, A. ramidus kaddaba, dating somewhere at 5.8 to 5.2 million years ago, contained one of the earliest discoveries of a hominid toe bone (pedal phalanx[7]). The evidence of the type of bone suggests that this creature was bipedal, though there's not enough fossil remains to make that a certainty. It is, however one of the divergent representatives of the early human form. Indeed, bipedalism is a distinguishing mark separating us from that long lineage that goes back to a common ancestor some 23 to 17 million years ago[8]. This creature lived, not in the dry landscape of modern-day Ethiopia, or even the golden savannah, but in a rich verdant landscape with lakes, swamps and a multitude of water sources[9].

"Lucy," the famed discovery along Olduvai Gorge in Tanzania, placed a milestone of fossil evidence for early human identification. Paleoanthropologist Donald Johanson identified an elbow bone sticking out of the ground in ground he knew to be at least three million years old. His trained eye told him the find would be significant, and indeed it was. The Australopithecus Afarensis, aka "Lucy," was found complete with metatarsal bones, pelvis, knee, and ankle bones. All of these demonstrated a bipedal creature standing about four and a half feet tall. Further evidence has been found about three million years ago by a set of footprints uncovered at Laetoli, Tanzania. These footprints not only identify bipedalism as the primary means of locomotion, they demonstrate the applied force of the foot. Indeed, these ancient ancestors walked not unlike we do today. Of course, other factors have to be taken into consideration when analyzing these prints (the grade upon which the prints were made, the density of the materials at the time of creation, and the creatures themselves.) However scientists have noted that the early humans walked with a distinct extended gait and extension of the foot not as great as present humans, indicating that they may have had a less developed "arch" of the foot[10] (medial longitudinal arch).

Another, earlier set of prints have been discovered in Trachilos, Crete. These prints also demonstrate bipedalism, and date this form of locomotion back to about 5.6 million years. The fact that these were not found in the "cradle of humanity" has tipped the apple cart for many anthropologists who have always used Africa as its basis for the dispersal of hominids. These prints show a large toe pointing forward - the distinct signature of human propulsion[11]. Some question whether there may be multiple paths toward what is present day vertical humanity, but the evidence indicates that the forward pointing toe is one of our distinguishing hallmarks and seems now to have existed in a variety of locations.

There have been other fragments of early hominids, but we cannot, with certainty, apply the kinetic model of bipedalism to these finds. Though traced back some six million years, they do not yield the evidence required to adequately build a case for bipedalism that far back.

"Of course, Ardipithecus ramidus and Orrorin tugenensis are strong candidates as well, but they also have too many anomalies to be undoubtedly declared the earliest hominin. All three candidates could be members of the various ape lineages that roamed before hominids began, instead of being the earliest member of the hominin lineage. For example, they could be a part of the common ancestry between humans and chimpanzees, rather than being the first human ancestors after the divergence between the human and chimpanzee lineages[12]."

Interestingly, the remains that do exist for Ardipithecus ramidus may indicate that the foot was built for grasping, like a primate.

The big toe is the beginning of the architectural change which altered our frame to become what it is now. Starting from the ground up, the toe led to changes in the knee, the pelvis, the back, the neck, and the head as we rose up from all fours. We are a rare species of mammal, unique our bipedal preference for everyday locomotion. I'm not talking about the occasional experiment when an orangutan reaches for high hanging fruit. Nor am I talking about the bonobo running toward its mother. In fact, the only other bipedal similarity to us is found in a variety of coastal wading birds.

It is those birds, like the flamingo, heron, egret, and cranes whom we most resemble when we traipse into the wet and muddy division between water and land. It is that which mostly distinguishes us from all other animals on the planet. Without the construction of our big toe and its alteration to support us in shallow waters, we would have never risen up to wade out even further, and forever alter our frame. And, in order to establish a coherent theory of our humanity, we must look to our construction, specifically that big toe, as the foundation for our bipedal gait.[13]

We take Darwin's theory of evolution for granted today, but in 1859 it was a radical idea. The evolution of bipedalism is understood as a part of this process. Natural selection isn't necessarily "survival of the fittest," rather it is selection which best fits competition for survival in the midst of limited resources. It is the reason the cuttle fish have perfected camouflage. It is the reason the hummingbird hawk moth has such a large proboscis. It is the reason poison ivy is - well - so annoying. Each modification through the process of natural selection improves one's chances as a species, which brings us back to the toe and bipedalism.

Following Darwin's theory, the reorientation of our toe, which was one element in the larger part of bipedalism should provide us with survival advantage. The results, however, seem contradictory to the event. As a two legged primate are we faster than our four legged cousins? No. Try chasing any critter from your yard. The four footed variety (unless you happen to be chasing a two toed sloth) will win every time. Our upright nature not only strips us of the advantage of speed, it exposes critical organs to prey. By standing erect, our center of gravity has changed such that the majority of the weight is above the center of gravity and not at its lowest point. This results in both osteoporosis and geriatric injuries. We, as unstable objects, are prone to tipping and falling (or slipping and tumbling with the hope that no one saw you). The location of the modern toe strips us of two grasping tools we once possessed. So why have we adapted to become the very thing that makes us slow, makes us vulnerable, gives us fallen arches, bad backs and neck strain? . Why then did our current, albeit imperfect, posture come into being at all? And, under what circumstances would the sum of those changes provide a Darwinian advantage?

A number of evolutionary hypotheses have arisen to explain our two footed stance. In reviewing these theories, keep the following in mind:

- Common Sense - What makes sense given the evidence we currently possess?
- Complete Picture - When developing the theory, is all of the current scientific evidence taken into account?
- Linear Progression - Does the theory fall in line with what happened before or after the timeline along the evolutionary scale?
- Stand the Test of Time - Given new techniques, discoveries, and methods in the scientific inquiry, does this theory still work?

Freeing the Hands

The theory is simple - We are bipedal to free our hands. Yet, in freeing our hands we have bound our feet. Where other primates have full dexterous advantage to grasp with four "hands" we are now limited to two. The toe is now impotent for grabbing, but much better adapted for supporting the weight of the body. Even this is imperfect as humans suffer significant toe injuries as a result of our stance.

Yet, from a common sense, this makes sense if the environment *requires* bipedalism. If we adopt the theory that our ancestors lived among the tall grasses of the savannah, then this makes no sense. Other animals gain advantage by remaining low to the ground, hidden amongst the thrushes and grasses. If we consider a scenario where we must carry objects *only with our hands* then this works. However, once we pick up an object with our hands, we are no limited in our mobility to a level platform. Tree climbing is no longer an option. The primate who is carrying food, and suddenly attacked will not be able to flee into the canopy for cover. However, in a coastal environment the feet would be embedded in a silty bottom. With movement of water, via a current born from a broke or river, the feet must remain placed and static while the hands are free to hunt or forage.

Low Hanging Fruit

Advancing the theory that our hands are freed, the theory of low hanging fruit is a natural extension. Our ancestors sought out fruit along the edges of the savannah. These young trees had limbs unable to support the full weight of the animal. The dental remains found in various sites suggest dental construction for consuming fruit. Today the Bwindi chimpanzees have a habit of standing erect in an attempt to grab fruit along the edges of the woodland environment[14].

This theory explains necessity for a static vertical posture, but the altered mechanics of the big toe isn't required for this transformation. Other than the necessity for standing on one's toes to get an additional inch, this theory has a few flaws.

First, this theory limits the environment in which our ancestors developed to a very narrow range. The total area where this is advantageous is very small. Considering the range where fossil evidence has been collected, the vertical stance for low hanging fruit seems less plausible. Low hanging fruit is mostly found along forest edges or open plains. Both of these are exposed areas. Though it doesn't discount the theory, it doesn't pan out when comparing the environmental remains found around archeological finds. Most bones have been found near dry beds that were once wet. This begs the question, were the bones found here because of the environmental conditions favored the preservation of the bones, or is this the location where hominids inhabited (this reducing the probability for the theory of low hanging fruit)

Second, early dental fossils, like today, indicate not only a diet of fruits and vegetables, but also a diet of meat. We were, in some ways, hunters as well as gatherers. This argument extends the range for food beyond the narrow small-treed forest edge.

The Throwing Theory

The basis of this theory is that we evolved into our erect stance because of the necessity to adapt to throw. Of late, this theory has become quite popular and the subject of many doctoral theses. Many research papers have been written comparing the superior throwing ability of humans over chimps and apes. However, for the same reason chimps are quadrupeds, their biological makeup makes them poor throwers. The chimpanzee hip is far forward of ours, preventing a natural vertical stance.

The anthropoligical premise is this: We adopted our upright stance because we want to throw things. Ponder this for a moment. You are standing upright. You are exposed. Some predator, much faster than you, sees you in mid salute as you cover your brow to examine the landscape. It starts charging for you. What are you going to do? Faced with, say a cheetah, would your first instinct be to reach down, pick up a rock and hurl it at the approaching animal?

Yeah - me neither. Even in a coordinated attack, the superior speed and agility of predators brings this theory into question.

The sister argument is that we gained an upright posture as part of our evolution for tool use. That seems to make more sense, unless you realize that, as far as we know, bipedalism predates the use of tools. The earliest spear is about half a million years old. This has some credibility, but again, there is no supporting evidence in the form of spears, mallets, or other devices that would support this line of inquiry dating back five or more million years.

Even accepting the theory, throwing requires incredible stress on the big toe. As the body pitches forward (in what is essentially an accelerated falling), the big toe bears the brunt of weight and flexing as the body is hurled forward. To be an effective thrower, the mechanics have to already be in place. The very scientists comparing our superior throwing to chimpanzees have, in effect, proven the demise of their own theory. The tradeoff between the survival of one who could throw moderately well, and surviving an attack is unlikely. We're not talking about six and a half foot humans, either. The early hominids come in at a little over four feet in height and are probably of slim build. This doesn't provide very much body torque, nor is there sufficient mass for offering a projectile against a predator.

Thermoregulation - Vertical posture and Hairlessness

The African savannah can be unforgiving in daytime swelter. This theory presents the idea that we have arrived at our upright stance to combat the savannah heat. To better understand the significance of this theory, we must couple the upright stance with hairlessness, perspiration, and our brain.

Imagine the smallish hominid as the sun bakes the savannah dry. While other animals seek shelter in the scrub and small trees, we are, for some reason, required to move in the scorching sun. By standing upright, only our head and shoulders bear the brunt of the direct sunlight - sort of. But evolution has favored us by stripping us of a protective covering of fur?

As is the bipedal hominid, the hairless primate is another oddity among savannah animals. These other animals are designed to allow for the increase in body temperature. But our brain is intolerant of such dramatic shifts in temperature. So we sweat. We do whatever we can in order to compensate to keep our brain cool.

Pete Wheeler, a physiologist at Liverpool John Moores University in England has studied this for years. For example, he took a one foot model of "Lucy" and moved it around a savannah, measuring its exposure to the sun. We are more efficient than quadrupeds in this environment according to Wheeler[15].

That works during the day - sort of. Animals, even our ancestors dating millions of years back, generally seek shade and comfort during a hot day. Couple that with the decreased efficiency in mechanical mobility and the advantage is negligible. The upright posture works at noon. But at night, this is a disadvantage. Other scientists who, in their wisdom, decided to research this theory at night, came to the conclusion that Lucy would have gotten cold.[16]

A study by Eugene Kayatkin demonstrates that an increase in temperature of the brain about two degrees celsius can have devastating effects. Though humans possess mechanisms to try and combat these effects, they may also work against the probability of survival (consider water loss, skin exposure, nutrients lost through sweating, etc.).

"While intense physical exercise in professional cyclists increased brain temperature to about 1°C or less, the same exercise conducted in water-impermeable cloth, which prevented proper heat dissipation, resulted in about 2.5 or 3°C (up to 40°C) increases (53) that clearly exceed 'normal' physiological limits.[17]"

Reaching for the Branch Theory

A recent theory regarding bipedalism surrounds the need to reach overhead for a branch. At the surface this makes sense. Go to any zoo and watch a primate reach up for branch. They assume a vertical stance,

This theory fails the common sense model gloriously. Returning attention back to the big toe, why would we lose our grasping capability?

Other Theories

Here are a few others. I'll allow you ponder as to why the big toe needed to be altered, and bipedalism became dominant:

- **The phallic theory** - Males created the bipedal stance to expose their genitals to females
- **The baby theory** - Females needed to breastfeed and carry their young
- **The fashion theory** - Primates look more handsome standing erect
- **The marathon theory** - The notion is that we ran down our prey not with speed, but with persistence. In the heat of the day, we would wear down our prey by running them down over the period of hours. Don't try this at home.

The Coastal Theory

The Coastal Theory, as it relates to bipedalism and the construction of the modern foot must be investigated with all available evidence, including but not solely looking at the hominid remains.

The general consensus in the scientific community is that bipedalism, for whatever reason, was a result of a changing ecology. The change in the savannah and its neighboring woodlands required our ancient lineage to rise up on two feet. The landscape of Africa was far different from what it is now. The verdant plains were lush and wet. Science proves that fact. The question is where we resided.

"We know that species as far apart in time as Sahelanthropus tchadensis 7m years ago and Homo erectus 2m years ago all lived in forested or open woodland environments….While some of these woods included wetland, this was just part of the mosaic of habitats that our ancestors learned to survive in, and there is absolutely no trace of a hominid ancestor as aquatic as that described by Hardy and Morgan."[18]

"The savanna scenario has lost some of its appeal since paleoenvironmental reconstructions started to show that the environmental setting has been more complex than was originally thought. Accordingly, more recent accounts describe the environment of early human ancestors as a mosaic of woodlands, savanna, and water bodies with considerable temporal fluctuations between climatically arid and wet periods[19]…"

The question is a matter of interpretation of what kind of aquatic lifestyle Morgan was presenting. To what degree had the early human evolved at, in, or near water? Does the dissenting scientific crowd mean that we as humans weren't around water? That seems unlikely.

One often overlooked fact with the discovery of Lucy was the fact that she was discovered amid artifacts of alligator and turtle eggs as well as crab claws. This implies that Lucy wasn't darting around the grassy knolls, but trolling through coastal waterways. The surrounding evidence of the fossil remains supports the Coastal Theory, not with anecdotal evidence, but with actual environmental clues as to the location of Lucy living (and dying) in an aquatic region.

If, in this scenario, we can imagine Lucy looking for shells in the water, or collecting crabs, the necessity for an upright posture is obvious. Even two feet of water, for such a small creature, would require the necessity for an upright posture. Consider, too, that along the coastal edge, Thrushes and other tall grasses would provide cover from predators. The water would make her buoyant. There would be a way to remain vertical, forage during the heat of the day, and remain cool in the process.

The earlier Ardipithecus, which exhibited more ape-like foot construction, still resided near lakes, streams and wetlands[20]. The discovery in Crete was found with water ripples in a shallow marine substrate[21]. Though these earlier discoveries are contested as to whether they draw a direct ancestral line to present day mankind, it highlights the environment where bipedal hominids developed the vertical stance.

Sound farfetched? Crab eating macaques, another genus of monkey, will wade into water, rising up on its hind legs, supported and buoyant in the water. The proboscis monkey has been found swimming more than a mile from shore. Bonobos, our closest genetic cousin, lives in the rainforest and swampy ares. They enjoy a diet including water lilies where they happily wade out for a morning snack. Some zoos create watery canals to separate orangutans and chimpanzees from the outside world. However, a few zoos were surprised when some of these animals taught themselves how to swim. Not only that, a few have been recorded diving down six feet[22].

Water provides the means by which the primate form can spend long periods in a vertical stance. For some macaques, this is part of their hunting environment where they spend a good deal of time. The transition from trees to coast changed the biological "tools" like the modern foot, to provide darwinian advantage.

This brings us back to the big toe. In water, the foot doesn't need to grab like it would in trees. It needs to elevate. It needs to balance. It needs to maintain a stance in current.

When wading into water, a grasping foot has only shifting sediments that would easily slip through its "hands". As a child I remember playing at the water's edge. I'd grasp at the sand as the water rolled and bit-by-bit took away what I held. A primate grasping foot would be no different.

As early hominids extended their search into deeper water, the modern day foot is perfect for anchoring (allowing the big toe to act to stabilize in shifting current and surge). It provides a means for extending when that one inch could mean drowning or not.

Other primates haven't the strength to extend onto their toes for any stretch of time. In the water, we humans excel at that operation, especially when a tall wave would otherwise put us under.

So the next time you rise up on your toes, consider our coastal ancestors and how they survived, thrived and eventually changed their locomotion as a result of a life in a semi-aquatic environment.

AMA TELLIN' EVERYBODY - THE ART OF BREATH HOLDING

Here on the ocean floor is the only independence. Here I am free.
— Capt. Nemo, 20,000 Leagues Under the Sea

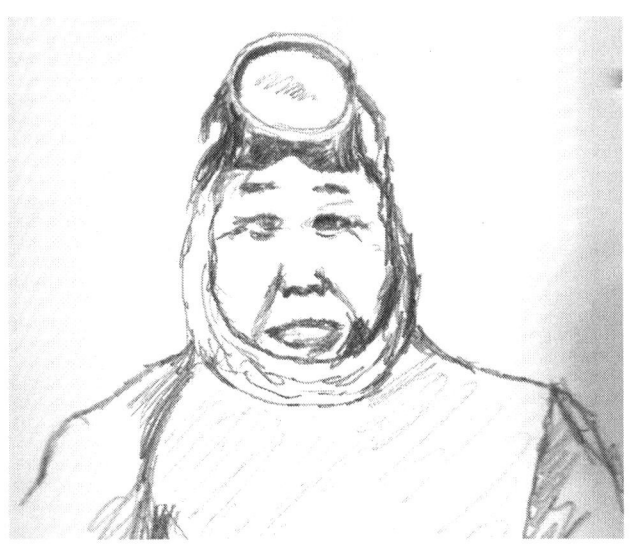

My friend Jon was on vacation visiting the American mecca of tropical bohemianism. It was one of those perfect mornings where the cumulus clouds billow high in vertical columns of pink and gold with a hazy purple line etching a band parallel to the horizon. You're never more than a few steps from the shore along the main drag of small shops that line the thoroughfare. At night the revelers are out in droves as the steel drums kick up their syncopated rhythms and unnamed dudes with guitars strum along to the reggae beat. It is the place to be both seen and unseen. The millionaires stroll in ratty clothes among those professors of cool, touting the qualities of the best sipping rum while on the foredeck of a custom yacht, young topless girls sway to hip hop beats.

But the morning begs a different crowd. Later in the day, those with a head recovering from a night of too much beer and spirits will seek out the small shanties to provide them with the life-altering balm of huevos with the hope of bringing them back to some semblance of normalcy. It's amazing what eggs and sunshine can do to remove the otherwise debilitating hangover. My research has included many of these locations, and I found a dash of hot sauce adds to the recuperative properties while masking the demon that invaded my breath.

A few natural holistic tea-totaling early risers see a different place. Many of the shops are closed until the cruise ship arrives, but there's one that opens at sunrise and closes by lunchtime. This is a specialty shop known around the island for the exotic specialties it offers. Only here can you get properly prepared sea urchin, baby octopus, and other obscure and exotic marine delicacies.

You'll know the ambiance of the store immediately. Everything is penned in poor English (read: "octopussies to for eight dollars"). Rotting wood crates are filled with the remains of creatures even a marine biologist would be hesitant to identify. Along both walls are lines of aquatic tanks all served with home-made aerators cut from PVC pipes to a loud motor somewhere behind the beach towel curtain that separates the seen from the unseen.

The clerks of the store are, in a word, unique. The youngest woman is seventy. The ages go up from there. These short and slight built women are still wearing their SCUBA wetsuit as they dole out the odd catch.

Ama divers have a rich and deep heritage. Once numbering in the thousands, these beautiful Japanese women clad in little more than a bikini bottom would plunge into the cold waters and stay submerged for minutes seeking out the edible riches of the deep. They seemed to define a type of glorious devolution – a return to the sea. Defying limits of both physics and physiology they became the mermaids of old. Long before the trendy apnea divers donned their monofin, these women of the sea passed on a tradition of techniques and secrets allowing them go deeper and swim longer than most thought possible.

The term "ama" (Haenyeo in Korean) actually translates to "sea women." Many have been known to take repetitive dives – as many as thirty or forty a day – to depths and for durations that would kill most of us. They've found a symbiosis with the sea. Those free-divers of old made themselves at home with the greater part of the wet world as they plunged easily to forty foot depths – the point where the body becomes negatively buoyant. This depth is often known as the "rapture of the deep." For free-divers (otherwise known as apnea-divers) it is a point of magic. Going deeper draws you ever farther from the meniscus between water and air and sends you downward with negative buoyancy. With no depth gauge or computer apparatus, these women would dive to where sunlight grew dim returning with baskets full of bounty from the depths.

Studies have shown in Bajau divers (aka sea nomads[23]) from Indonesia a marked alteration in their genetic makeup, providing physiological alterations - specifically a larger spleen. What role does the large spleen play in the role of breath holding? When the Bajau dive in cold water they hold their breath until the mind screams for air. Only then does the spleen releases a pool of blood rich in oxygen that prolongs the dive. Some Bajau divers are known to regularly hold their breath for as much as thirteen minutes.[24]

Archaeologist have discovered artifacts and bones from Chilean divers, thousands of years old, whose bones clearly demonstrate "surfer's ear" - the abnormal bone growth (exostosis) that results from repetitive deep diving in cold water.[25]

Today freedivers and SCUBA enthusiasts around the world have taken these dormant and all but atrophied abilities and resurrected them. What has been practiced around the world for centuries seems to have been forgotten in modern civilization. The advent of SCUBA created the necessity for military research with the naval "frogmen" in the fifties and sixties. This included those wearing apparatus to breath underwater or those practicing extreme breath holding while diving beyond recreational depths.

So, what happens to the lungs when freediving that would cause concern? In order to answer that question it is important to understand the scientific principle coined by the seventeenth century scientist Robert Boyle. Boyle's law states:

pressure and volume for a gas are inversely proportional

To better understand what happens to the lungs, let's conduct a little "thought experiment."

- Get in a boat (at sea level) - sail out into the ocean and set anchor
- Blow up a balloon until it is nearly ready to pop.
- With a good breath of air dive down about thirty-three feet (ten meters)
- Look at the balloon which is now half its original size

Why is the balloon half its size? By diving down to that depth you are at two atmospheres of pressure (at sea level on the boat you were experiencing one atmosphere of air - at thirty-three feet underwater you have one atmosphere of air plus one atmosphere of water). What has happened to the balloon has also happened to your lungs. Dive down to about 66 feet (20 meters) and your lungs are a quarter of their size.

Navy scientists feared that at greater depths, the thorax would collapse, the rib cage would fail and the human body would implode due to the pressure on the body.

But this was not the case.

Physiologist Per Scholander wondered the same fate as he studied other mammals who regularly hunted at far greater depths. Why did the depths not seal their fate? His study of Weddel seals revealed a shift in the blood to different parts of the body. Could the same happen to humans? His studies confirmed that the blood in humans, like their aquatic counterparts, left their extremities and headed toward the core, protecting the lungs.

Imagine our balloon experiment with the exception that with the balloon at depth you were able to inject fluid into walls of the balloon so that the overall area of the balloon didn't shrink. The fluid compensates for reduction in airspace, retaining the original architecture at the surface. This is what happens to us when we dive.

Scholander coined the term "Master Switch of Life" though the more common term is the mammalian dive reflex. This little understood transformation of the the body as one dives to greater depths is still little understood in the volumes of science on the human body.

As elegant as the lungs responding to being deeply immersed, it is but one of many things that occur during the mammalian dive reflex to keep us alive. Not only does our blood compensate by centering itself at the core, our heart rate slows down - beyond what many doctors ever considered possible. At a mere two atmospheres (66 feet or 20 meters) our hearts beat at half their normal rate. For those free divers who have broken the magical 100 meters (300 feet):

"The organs collapse. The heart beats at a quarter of its normal rate, slower than the rate of a person in a coma. Senses disappear. The brain enters a dream state."[26]

In that sense, the alteration of the physiology and psychology of breath holding are tightly intertwined.

At the core of this mammalian reflex is the result of an interesting evolutionary detour. As unique as bipedalism, hairlessness, tears, among other traits defining "human," the architecture of our throat and our ability to speak harkens to a coastal evolution where breath holding was a necessity. Of course, we're not the only animal who uses sound to communicate. But as primates go, our pharnyx, velum and epiglottis (that thing that causes hiccups) differ radically from our primate siblings. Our airway is a more open highway with connections between our nasal passages and throat to our lungs as well as our stomach.

Other primates have two passages - one from the sinus to the lungs and another from the mouth to the stomach. In fact, infants share the trait of our ancestors and other primates allowing them to suckle their mothers while simultaneously breathing. Only later does the dual pathway open up to become one - which can cause us to choke on our food.

What prevents the food from going into our lungs is the epiglottis which covers the airway to the lungs. This fold of muscle covering the airway allows us to enter the water with our mouth open without filling our lungs with fluid. By the way - hiccups are an involuntary muscle contraction of the epiglottis. The popping noise is the involuntary and rapid opening and closing of the airway[27].

Additionally we have a flap in the back of our mouth leading to the nasal passages which we can use to prevent water entering our sinuses. Many who claim they can't put their head in the water because water goes up their nose don't or can't properly seal with that flap.

It is difficult to identify the exact period we transitioned from our primate "dual highway" air passage to what we have now. The soft tissues and tendons do not stand the test of time. Only guesswork based on the shape of the cranium, the elongated new and modification in posture give us a glimpse of the change. There is also the difficulty with skeletal morphology (comparing the craniums from like ages) because of the uniqueness found in the traits of the surviving record of humanity. Just as we are different, the skeletal remains provide only a comparative glimpse over a span of millions of years.

Scant attention is given to the study of breath holding in early hominids. However, we can infer the origins of breath holding from scientists curious about the origins of human speech. Speech is nothing more than an organized breath hold with a calculated exhale. When we articulate a sentence we calculate the amount of air we'll need to get from beginning to end. What seems like a preposterously simple activity takes a great deal of planning and forethought. There is pressure differential as we raise and lower both pitch and volume. We exert more air on words we wish to emphasize. A whisper is a minuet compared to the coaches roar above the cheering crowd. Yet, both require a great deal of effort and understanding of the breathing technique.

It is through the research of speech we come to the conundrum of the chicken and the egg. The greater attention toward the human distinction of bipedalism reigns over the lesser uniqueness of the lower position of the larynx and the enlarged sinus cavities making complex articulated speech possible. But which came first?

Scientists argue that the changes in the cranium and its position on the body forced a restructuring of the human nasal passage. The massive sinuses and size and shape of the nose was a result of a necessity to "air-condition" the air with the heating of the planet. What is fascinating is that primates actually have an easier time breathing than we do. The septum and reduced nasal passages actually restrict our ability to effectively breath through our nose[28].

But another explanation is at hand. If one considers aquatic hunters, the restricted passage of fluid up the nose is an evolutionary benefit and not a hindrance. The larger maxillary sinuses provide a greater well of air readily absorbed through the membranes that feed the brain. Water that may get into the nose can pool there rather than enter the lungs. Anyone who has ever had one of those pressure filled, heavy draining colds knows that the mucus pools rather than drains and that our human plumbing is geared toward retaining fluids in the smaller channels of our sinuses. This reduces both the choking and gag reflex one would expect with a rush of fluid toward the lungs.

Interconnected with the sinuses are the vestibular system which is an important adjunct to the ear. This system helps us orient up, down and our position in space. This becomes increasingly important with hunting, running, or circumstances where we need to fixate on an object while in motion. Studies have correlated the size of the semi-circular canals which allow balance to the size of the organism. One exception, however, is humans. Our canals are much larger. This has been attributed to the need for greater stimuli as a bipedal animal. However, in water, this provides exceptional advantage where hunting takes on a third dimension. These canals help orient humans with roll, pitch and yaw. So this begs the question of advantageous adaptation. Was this a result of our bipedal nature, or a consequence of our aquatic hunting practices?

If our ancestors were aquatic hunters, surfers, and foragers, then how deep did they dive? Did the early hominids have to equalize when they dove? The answers to that depend largely on location and environmental conditions. Many primates enter water that exceeds their height, but have few places where they could dive to a great depth. Wading and foraging, taking advantage of a wet passage, or entering water for safety are a few conditions where primates enter water today. A few examples of primates in captivity near water sources prove that primates can and will dive.

Cooper, a chimpanzee was filmed doing a rather impressive breaststroke in Missouri. An orangutan name Suriya has developed a swimming technique all her own. They have been found diving down six feet (two meters)[29]. The proboscis monkey has been found swimming a mile from the nearest island.

These are learned responses and not innate to the primate. This is also true of humans. There is a myth that infants have an innate ability to swim. This is just not true. Elaine Morgan and the Aquatic Ape have promoted this myth, sadly to find many of the "water babies" classes. Many have drowned as a result of this fallacy.

So how are dolphins, manatees, whales and other ocean faring mammals able to remain submerged for so long? If you've ever had the misfortune to see a dismembered whale or dolphin you'll note that their muscle tissue is extremely dark, looking more like a rich red liver than ordinary muscle. There is a protein called myoglobin found in muscle and skeletal tissues of all animals, but is particularly rich in aquatic species. This substance (a cousin to hemoglobin) stores oxygen in the muscle tissue where it is consumed close to the source as the creature dives deep.

Our primary transference of oxygen in the body is through hemoglobin in the red blood cells. That's why a number of top athletes train at high altitudes. The body is forced to make more red blood cells. Less trustworthy athletes regularly "donate" blood to themselves - storing blood, spinning it down for the red blood cells, and then re-injecting it into themselves so they have a higher proportion of red blood cells than their competitor. This is known as blood doping and can have significant ill affects. The blood is thicker so it has a harder time moving through the body. The body can turn on itself attacking the aged blood resulting in blood clots. The system also has to take care of cleaning out the extra stuff so there is a toll on kidneys and liver function.

Where hemoglobin tracks through the circulatory system, myoglobin is static - remaining in the muscle and bones until it is needed during a hypoxic situation. Humans have trace amounts of myoglobin in their body. There is not enough to keep a human alive submerged for any great period of time. Interestingly, during muscle injury, the human body produces extra myoglobin at the effected area which can be spotted and accounted with a corresponding rise in the circulatory system.

Free divers have questioned the scientific community about ways to increase the levels of myoglobin in their muscle tissue. From increased selenium and Vitamin E to altitude training, they've searched for the edge to make them more like their cetacean brethren to no avail. Research has found that these methods do not significantly increase the levels of myoglobin in the muscle tissue.[30] For humans, at least, hemoglobin is the only answer for oxygen transference. Some free divers have "breathed up" - that is - gone through a practice of breathing exercises to super oxygenate their bodies. When doctors have taken a blood sample, it is rich and dark as one would expect. After the free divers submerge and perform their breath holding sport, the doctors, on a subsequent sample, have found the blood pinkish blue and severely lacking in color because of the lack of oxygen.

Were the early walkers on land actually hunters of a different kind?

At some point our ancestors waded into the waters and evolution followed them. Today, we have reinvested in that watery quest, now for entertainment rather than survival. But deep in the back of our mind, is the echo of some primordial calling to plumb the depths of the darkest blue.

Greek sponge divers would compete to see who could hold their breath the longest. Navy divers would brandish their machismo by diving off their boats holding ballast. From kids in public pools to hippie surfers, groups started to form around breath holding competitions that seemed as elusive to find as the "locals only" surf competitions.

As a kid, I lived near a public pool. I spent the greater part of my summer on a beach blanket, reading science fiction and looking at the girls. But when Ray Metzer and Bill Forder showed up, it was all about breath holding. We'd go to the deep and "breathe up." What was interesting is that we had no formal training, nor did we know about the free diving community at the time. It was an extraordinary experience as we attempted to swim the length of the pool. When we accomplished that feat, we challenged ourselves to swim an entire lap in a breath. Those that knew us would gather at the end of the pool to see who would go the farthest. We'd dive down to the deep and hang on to the grate to keep from surfacing. There was pain associated with this experience, but there was also a great peace about it. We'd get kicked out for the day by the life guards (who knew us by name) when they came to rescue us after we'd been down for what seemed like too long. To this day, I still practice (albeit with spotters and a little more of a conservative approach) breath holding.

In the 1970's a disorganized tribe free divers attempted to push the boundaries science had imposed. The competition between Jacques Mayol and his friend Enzo Maiorca redefined breath holding techniques. Other divers primarily competed by hyperventilating before their attempt, often with disastrous results. Unlike the Ama divers, these competitors would "over breathe" - building up the oxygen in the blood stream. The reduction of carbon dioxide fools the brain into thinking the diver has plenty of oxygen - that is until they, often without warning, passed out. A byproduct of hyperventilation is an increased heart rate. This results in a quicker consumption of oxygen.

Mayol challenged this technique, looking instead at relaxation and mental techniques over hyperventilation. Adopting yogic practices Mayol developed new training methods and on November 23, 1976 broke the 100 meter dive mark descending to 105 meters. He worked on techniques to expand his ribcage and modified breathing techniques to maximize air consumption while reducing physiological and psychological anxiety. The larger space and "lung-packing" provide additional oxygen while providing additional volume for storing carbon dioxide.

These techniques were made popular in a 1988 film *The Big Blue* which fictionalized the life of Mayol and Maiorca. When I saw this film, it changed my life. I realized there were others out there doing what I did locally. There was a community who experienced what I experienced. I wasn't alone, a trio of boys who discovered something special. Others were called to the sea to challenge what humans had deemed impossible. The film also showed how many died or were critically injured in their attempts. It also highlighted the need for better organization of this life-threatening sport. In 1992 AIDA (International Association for the Development of Apnea) was formed. This organization combined the competitive, instructional and scientific aspects of the sport. Today the sport comprises a number of competitions including:

No Limit (NLT)

The freediver descends with the help of a ballast weight and ascends via a method of his choice. No limits is the deepest depth discipline. The athlete descends with a sled and ascends with a balloon, a diving suit, or a vest with inflatable compartments, or whatever other means. NLT attempts are currently not sanctioned by AIDA International.

Many have died as a result of mechanical malfunction. Some have deemed the home-made sleds "express way to Davey Jones's Locker." Should the machine malfunction, there is little recourse for a proper ascent. The speed with which the machine descends can have devastating results on any part of the body that requires equalization.

Variable Weight (VWT)

The freediver descends with the help of a ballast weight and ascends using his own power: arms and/or legs, either by pulling or not pulling on the rope. Variable weight is one of the two depth disciplines which employ the use of a sled to descend in the water. With old-style sleds, the athlete descended "head first" as seen in Luc Besson's famous movie "Le Grand Bleu", but new generation sleds are "feet first" which allows for easier equalization for the athlete. Variable weight is only done as a record attempt and is not a competition discipline.

Constant Weight (CWT)

The freediver descends and ascends with the use of fins/monofin and/or with the use of his arms. Pulling on the rope or changing his ballast will result in disqualification; only a single hold of the rope is allowed in order to turn and stop the descent and start the ascent. Constant weight is the most widely practiced and known sportive depth discipline of freediving due to the specific fins or monofins used in it. Constant weight is one of the three disciplines included for international team competitions along with Static apnea and Dynamic with fins.

The concern with the constant weight system is the change in buoyancy as one descends. A neutrally buoyant individual at the surface quickly becomes negatively buoyant. At depth, when the individual starts to become oxygen starved, the swimmer must exert themselves more during this period, to rise to the surface.

Constant Weight Without Fins (CNF)

The freediver descends and ascends underwater using a variation of breaststroke swimming stroke without the use of propulsion equipment and without pulling on the rope. Constant weight without fins is the most difficult sportive depth discipline because it requires the most strength and the diver is unaided by fins. CNF exemplifies perfect coordination between propulsing movements, equalization, technique and buoyancy.

Free Immersion (FIM)

The freediver dives under water without the use of propulsion equipment (fins) but uses the rope to pull to descend and ascend. Free immersion is the most relaxing discipline and is used as a training tool to learn equalisaton techniques. Athletes may experience the most enjoyable sensations in FIM because of the speed of the water over the body and the power of each pull on the rope as the only means of propulsion.

Dynamic With Fins (DYN)

DYN is most often a pool discipline in which the freediver travels in a horizontal position underwater attempting to cover the greatest possible distance. Any propulsion aids, other than fins or a monofin and swimming movements with the arms, are prohibited. Dynamic with fins is the most common of the horizontal distance disciplines because of the specific means of propulsion; long fins or monofin. For a performance to be officially recognized there are minimum depth and length requirements for pools. Pool competitions most often comprise a performance in DYN, DNF, and STA but some competitions are a combinination of DYN and STA.

Dynamic Without Fins (DNF)

The freediver travels in a horizontal position underwater attempting to cover the greatest possible distance using a modified breaststoke. Propulsion aids of any sort are prohibited. DYN requires good technique, relaxation, and a long breath hold in order to achieve the greatest distance. The minimum pool standards are the same as for DYN.

This is much safer than diving to depth as you do not risk some of the physiological issues associated with diving deep. You also reduce error in judgement as a result of narcosis.

Static Apnea (STA)

The freediver holds his breath for as long as possible with his nose and mouth immersed while floating on the surface of the water or standing on the bottom of a pool. Static apnea is the only discipline based on time of breathhold and not distance. It is one of the three disciplines included in Team World Championships along with CWT and DYN . Performances are recognized in both pool or open water (sea, lake, river, etc).

A number of elements of the free diving experience make this challenge deadly. But number one on the list is shallow water blackout. Those competitions where divers descend and ascend to the surface risk the demon of shallow water blackout.

To understand shallow water blackout we need to revisit that balloon that we took down to depth in a previous example. Imagine filling, packing, and cramming as much air into your lungs just before you hold your breath and dive down. There, at the surface your body feels uncomfortable because you have expanded your lungs to their limit.

When you hold your breath, the reduction in oxygen in the lungs is not the trigger that makes you want to take a breath. Rather it is the body's "sensor" in the brain that reacts to levels of CO_2.

You dive down to thirty-three feet. Now your lungs are half their size. No longer do you feel the discomfort of lung over expansion. Not only that, but the density of the air has doubled. Note - the size has decreased - but the number of air molecules has not changed. Therefore, with the decrease in volume, the number of molecules hitting the lung has doubled. You feel relaxed and can hold your breath without exertion. Add to that you have a rush as your body becomes more efficient with the oxygen traveling to your brain.

Your body is going through its mammalian reflex. The blood is starting to shift away from the extremities. You start to get the diver's "high" as chemicals are released to the brain, signaling the ancient return to the sea.

You now dive to sixty-six feet. You are now barely negatively buoyant so you are no longer struggling from rising to the surface. Where you had to kick to go down, the rapture of the deep has taken over and you relax as you feel the water course across your face. Depending on where you dive, it may be difficult to determine your depth. If you are in the open ocean, you have no visual clues to identify your frame of reference. Sometimes you merely have a down-line to follow with the odd marker to indicate depth.

Your brain is slowing down. You realize that even thinking takes energy. Some dive with their eyes closed. During one competition, a diver was performing his dive. Minutes later the spotter surfaced signaling the diver was lost. He never reached his intended depth. A search was started. After Thirteen minutes he surfaced - on another down-line! He'd gotten lost, swam around at depth and finally found a line to guide his ascent.

For really deep dives during breath holding it is important to understand the mechanism of oxygen transfer from the lungs to the bloodstream. The bronchial tubes that start from the trachea divide into smaller and smaller "pipes" (bronchiole) ending with microscopic sacs with very thin walls. These sacs are called alveoli. Surrounded by capillaries, they are so thin that gases (oxygen carbon dioxide, etc) can pass between these walls.without the blood moving into the lungs. The process for this transfer is called diffusion. Gases move from an area of high partial pressure to an area of low partial pressure. There is more oxygen along the outer wall of the alveoli than inside the moving capillaries, so oxygen diffuses through the alveoli wall. Conversely carbon dioxide has a higher concentration in the bloodstream so moves from the bloodstream to the lungs.

At extreme depths the process alters. Massive pressure forcing the lung to a fraction of its size blood can enter the alveoli sacs, filling them with blood. This prevents the sacs from collapsing while still offering the necessary gas exchange. It also explains why divers surface coughing blood or having blood about their nose and mouth. This is particularly scary because the symptoms are identical to pulmonary edema, which can lead to respiratory failure.

Divers who hold their breath for extended periods force back the natural urge to breathe through sheer act of will. Many go through a period of convulsion where the body contracts in an involuntary rehearsal for a real breath. Apnea divers call this the "samba." The body - realizing a desperation for new air isn't forthcoming will release oxygen rich blood from the spleen, giving the brain a dose of what it needs.

The heart slows to a rhythm not thought possible to sustain life. Many free divers' heart rates get down to about fourteen beats a minute. Some have been recorded as low as seven. Scientists say this rate is too slow to support consciousness.[31]

You are in a state of euphoria because of narcosis and the body's natural mechanisms for dealing with oxygen starvation. The urge to breathe has just started to tap you on the shoulder. You have consumed half your oxygen, but you barely notice because you have four times the oxygen hitting the walls of your - now tiny- lungs. Some blood and other fluids have started invading the alveoli, those areas that help transmit the oxygen to the bloodstream. This is due to the pressure and the bodies natural mechanism to protect the lungs from collapsing.

You begin to surface. Your lungs are expanding and suddenly you feel oxygen starved like a man who hasn't eaten in days. The brain - once content and euphoric is now tortured to get to the surface. In some cases, divers force themselves to hallucinate to compensate for the immense pain that takes over the senses. Why are you in so much pain? When you were at depth the oxygen molecules bounced off your lungs, now they have farther to travel - those that are left. Those that hit the wall may not translate to the bloodstream because of the fluid that lines your cell wall of the lungs.

You are fighting the urge to breath as lactic acid courses through your muscles, causing them to constrict as they fight for air. Luckily your training has told you to relax through that pain. You see the surface, but your lungs cannot supply the oxygen to the brain - the largest and hungriest of oxygen organs in your body. It cannot sustain consciousness. You look up but only see a pinprick of light. All else around your field of vision is black. So - with only a few feet to the surface, you shut down.

This is shallow water blackout.

Some, who are lucky, still maintain some primitive sense of consciousness even while shut down. They are aware enough to know not to breathe as they rise to the surface. Those lucky enough to get there will find spotters, and assistants coming to their aid.

The body - even at the surface - refuses to breathe. One technique, known to the dive community but still perplexing scientists, to get the diver to breathe is to blow across their face[32]. The "wind" causes the senses of the face to detect air and sometimes the free diver will come back with a quick exhalation and series of short breaths. These are the lucky ones.

Still, with the dangers of this game, the free diving community grows every year. The sport has increased in popularity to the point that other agencies, specifically those in the SCUBA society (PADI, SSI, etc.) have joined the ranks of AIDA in providing free diving training. Nonetheless breath holding is crucial to understanding our watery evolution. One side of the story defines our uniqueness among primates in our capacity to temporarily tour the other seventy odd percent of our planet. Yet, apnea diving has uncovered the dangers and limits which separate us from the marine mammals who seem unencumbered by the physics of going deep.

There is a part of our evolution that calls us back to the depths. Perhaps it is the company I keep, but it is rare the person who hasn't felt some compelling unexplained magnet that draws them toward the ocean. According to a 2010 NOAA survey, 39% of the nation's population lives within an easy drive to the beach. In fact, it is six times greater than other counties further from the coast[33].

Though the story of diving and breath holding as part of our evolution is still silent, it is interesting to ponder that our mammalian reflex as well as our capacity and physiology to be able to hold our breath makes us unique among primates. Our very language is dependent upon our ability to hold our breath.

If there was ever a punch line in this rhapsodic look at breath holding we need to return to the ama women found on this small island in the United States. The lines form early for their latest catch. Men and women who have chosen to become pescatarians (fish and veggie eaters) seek out their odd choice for a day's meal. Many of the people define cool - they have the right sunglasses with the strap and cotton outfit that came from guaranteed fair-trade online catalogs. Each owns a signed Wyland print mounted somewhere on the wall of their apartment that overlooks the beach.

My friend, Jon laughs silently as he looks on at these customers. You see – Jon understands a host of Asian languages. As he peruses the wares he listens to the ama divers. Unknown to the "natural chic" clientele these women never touched the water that glistened at the crack of dawn. They haven't been diving in years. These "ama" divers imported the aquatic oddities from overseas. Perhaps what Biff the naturalist has been consuming was brought in from Fukushima. Who knows? These old women in their wet suits thaw out the stuff in a microwave before opening their doors. For them it's a matter of commerce. The tourists buy what they want. They get the street cred for buying from a sweet old lady who can speak only pigeon English.

I can only laugh having been on the inside of the joke.

WET GORILLAS

An American monkey, after getting drunk on brandy, would never touch it again, and thus is much wiser than most men.
— Charles Darwin

As the afternoon wore on and others succumbed to the doldrums of the flat landscape, the "PADI Wagon" SCUBA bus grew silent save for a few snores and heavy breathers. In the quiet and my own post-SCUBA buzz I started daydreaming about my past. I recalled those savannah dioramas from the Smithsonian. I was contemplating man's evolution and the possibility of the Coastal Evolution Theory as a mishmash of free associations rambled through my head. At one point I thought of the small hairy creatures and made the leap to one of my favorite movies - "2001: A Space Odyssey." For over a half hour I watched a wordless cinematic magnum opus of early hominids fighting around a small watering hole. The accuracy of the hominids were so accurate that the Emmy board of that day never offered the director, Stanley Kubrick or his staff, best costume because they didn't know they were actors in a suit. On the silver screen I witnessed the birth of intelligent humans - fostered by a black obelisk that rang out in a chorus of random voices. It was a radical alternative to the

orderly arrows that pointed ever upward along the case of skulls in the Smithsonian.

The theory of evolution discovered in the engine of natural selection was, at the point of its creation, a driving force that radically challenged humanity's notion of perfection through divine providence of intelligent design. To date there are still those who believe in four thousand year old creation-to-completion myth. Both Darwin and Wallace (who proposed the same theory and has an interesting history all his own) met with strong opposition - even in the scientific community.

Darwin's definition of "natural selection" is as follows:

"How will the struggle for existence, discussed too briefly in the last chapter, act in regard to variation? Can the principle of selection, which we have seen is so potent in the hands of man, apply in nature? I think we shall see that it can act most effectually. Let it be borne in mind in what an endless number of strange peculiarities our domestic productions, and, in a lesser degree, those under nature, vary; and how strong the hereditary tendency is. Under domestication, it may be truly said that the, whole organisation becomes in some degree plastic. Let it be borne in mind how infinitely complex and close-fitting are the mutual relations of all organic beings to each other and to their physical conditions of life. Can it, then, be thought improbable, seeing that variations useful to man have undoubtedly "occurred, that other variations useful in some way to each being in the great and complex battle of life, should sometimes occur in the course of thousands of generations? If such do occur, can we doubt (remembering that many more individuals are born than can possibly survive) that individuals having any advantage, however slight, over others, would have the best chance of surviving and of procreating their kind? On the other hand, we may feel sure that any variation in the least degree injurious would be rigidly destroyed. This preservation of favourable variations and the rejection of injurious variations, I call *Natural Selection*[34]."

From beetles to birds Darwin unwound a once radical notion about evolution. The basic premise is that those who are able to adapt and survive are those who are available to procreate. Interestingly he uses climate change as a means to describe the "adapt or die" method for species dealing with an altered ecosystem. He uses natural selection as an avenue for mutation and thus the opportunity for biodiversity as a means for survival and ongoing perfection of a species. This correlates to the lineage of those skulls rising ever and changing. The need for a square jaw and larger brain case was because we were able to adapt to meet a need. Our upright posture bought us advantage. Therefore the human mutation over the millennia provided constant improvement as our environment, predators, and surroundings changed.

My science books from high school explained how the evolution of our upright posture gave us a greater view of the savannah. Our hands, in particular the all-important opposable thumb, provided a way for us to make tools. Our tool making ability occurred simultaneously with the ever increasing size of our brain.

Darwin was quick to identify that natural selection was the primary, though not the only means, to explain the mechanisms of evolution. The distinction between evolution and its mechanisms is important. One who only looks at "survival of the fittest" fails to explain the variation and rise of some of the most beautiful yet fragile life on earth. It fails to explain the radical turns in species and adaptations that make no sense. The lengthy and general rise of natural selection fails to explain the tight evolutionary corners that can be turned in a single generation.

The drama of evolution as played out through the paleontological theater required that its priests and cardinals uphold a doctrine showing its age. Not that evolution as a scientific model was wrong, but that there were amendments as a result of new evidence requiring contortions to support Darwin's theory.

As a human being with a stiff back and sore feet, I didn't feel like I was the king of the "fittest." The interplay of ecosystems allowed for the weak to become dominant. Perhaps Darwin was only partially right. Perhaps the pieces to the puzzle were being forced together to build a picture the scientists wanted rather than reassembling the evidence in other ways to explain the mystery of the imperfect human. Darwin added to the equation - but I felt the equation incomplete. Perhaps the scientists were wrong.

Upon my return to civilization from my long dive weekend, my initial research revealed that I wasn't the only one who questioned this theory and its later amendments.

Consider a single trait that separated us from our primate brethren - Hairlessness (Or better coined - furlessness). It would seem a trivial matter, but one of extreme importance to identify the advantage we gain by losing our fur - and gaining hair. I'm not sure about you, but I've never seen a monkey having to shave. I've never witnessed an orangutan requiring a hair cut. That is because they have fur - we have hair. As a balding man - I still have to get my hair cut lest I look like some creepy gray-haired Bozo-the-Clown. The academics in archaeology talk about hairlessness as a means to cool down. I've never seen a hairless camel. There aren't a plethora of skin exposed animals in the savannah.

The Parasite Theory - Dating back to the date of Darwin, the loss of fur was to prevent the accumulation of ticks and other parasites. Of course this fails to explain the fact that other animals, living in the exact same circumstances retained their fur.[35] However, if you add to that the fact that the hair/fur might be wet, then the theory holds as a contender along with the Coastal Theory.

The Bipediality Hypothesis - The bipediality hypothesis put forth by Wheeler (1984) argued that the reduction of body hair was made possible by the lower direct solar radiation fluxes incident upon a bipedal mammal, which also explains the absence of this characteristic among savannah quadrupeds[36]. Comparative anatomy of other mammals (like a kangaroo) who maintain a somewhat vertical stance, still retain a thick coat.

The Allometry Hypothesis - The term allometry refers to the fact that as species become larger in the course of evolution, not all organs of their bodies increase in the same ratio as their overall body size and mass. We have hair but it appears smaller because we are larger. Given the nearly doubling in size, the hair on our body still does not compensate for the amount lost, nor does it account for the locations (like the shoulders) where a protective coat would be advantageous.

The Hunting Hypothesis - vegetarian primates do not need to move very fast; a carnivorous primate, on the other hand, would get hot while chasing its prey, and losing its hair would enable it to cool down. This presupposes that the early hominid would be hunting when the sun is at is apex. The cooling of the body would have to be combined with the ability to sweat. Therefore, the efficient hunter would have to be constantly seeking a water source, given that the hairless body, mid-day hunting, and bipedal creature would be profusely perspiring. Again, if one wants to adopt this theory, it would have to be in light of a hominid within close proximity to abundant water.

The Clothing Hypothesis - It has been suggested that clothing made hairlessness inconsequential and that hair reduction provided no intrinsic advantage. Clothing and tools go hand-in-hand in our history, but bipedalism and hairlessness predate the use of either.

The Neoteny Hypothesis - Neoteny is the retention of juvenile physical characteristics in the adult individual[37]. If you look at an early infant chimpanzee, the hairlessness and general physique looks more human than when they are full-grown. The premise is that we turned off our genetic construction for a hairy human as part of a genetic mutation which rippled through our evolutionary tree to make us the infantile representation of the hairless human. When we talk about genetic radiation (the dispersal variety, not the nuclear waste variety) it is the mutation which generates advantage, or is at best benign, that moves outward in ever increasing circles. Hairlessness, as a matter of accident provides no benefit (as declared in the arguments above) and actually hinders our ability to protect our bodies from exposure to the sun and insulate our body at night.

The conundrum of hairlessness has plagued paleontologists and speculators of the human condition for centuries. With little forensic evidence, one can only ponder "what if" games. The above mentioned theories are but a few of the more popular ideas floated through the ages. We can dismiss the Allometry hypothesis outright because we have as many hair follicles per square centimeter as our furry brothers and sisters in the primate world. What sets us apart is the fact that our "fur" is so light and small, that it is all but invisible. Some is so short that it doesn't even break the surface of our skin.

We can break "hair" into three basic categories. These categories are not as much based on physiology as they are on use.

First is whiskers. One trivia fact that I find personally fascinating is that the length of a cat's whiskers is equal to its width. The whiskers have a set of sensitive nerve bundles associated with them, allowing the cat to detect objects in the dark as it brushes along its course. Manatees also have whiskers, as do horses, rats and a host of other animals.

"Whiskers are vibaissae, keratin filaments that grow out of different follicles than hair. Whisker follicles are much deeper than hair follicles, and are surrounded by pockets of blood that amplify vibrations to better communicate information to the nerve cells beside the follicles. You may have noticed when looking at your cat that there are 2 kinds of whiskers, long and short. Long whiskers are macrovibrissae and can be moved voluntarily. Animals use these to sweep areas (called whisking) to navigate spaces and generally feel the world. Short whiskers are microvibrissae, and they cannot be moved voluntarily. These are used specifically for object recognition, whether it's your cat's favourite toy or your hand[38]."

Second is hair. Hair and fur are, for the most part, identical save for the length and duration of growth. Human hair comes in two flavors: terminal hair, which grows on the scalp, eyebrows and eyelashes, and vellus hair, which is found everywhere else.

"The 'anagen' period of the cycle is the phase of constant hair growth; the 'catagen' phase is a transitional time when the body tells the hair to begin to stop growing, shrinking the strands themselves and cutting the roots of the hairs so new strands won't be produced; and the 'telogen' phase is when the hair follicle rests and no new growth occurs. This leads to the 'exogen' phase, in which the hair falls out to start the cycle over again. So in humans, hair will stop growing after the cycle runs through each phase. It's just that the cycle in you as a human is longer than, say, Fido's hair growth cycle. The anagen period of active growth on a human scalp could run anywhere from 2 to 7 years (taking into consideration other factors such as baldness), while the telogen period in which the hair on your scalp is dormant could go up to 100 days[39]."

Another element of human hair is that, along with the stump-tailed macaque, we are the only ones who share a receding hairline! Most scientific rigor toward the receding hairline has little to do with our evolutionary distinction and has more to do with how to retard the process. Hair length is another question to be answered and some have speculated that the long hair requiring cutting is a later genetic addition. African Americans, with shorter, curly hair, it is assumed, would be representative of the area and type found five million years ago[40]. The question is, if baldness is a more recent phenomenon, then it may not necessarily play into the question of the Coastal Evolutionary Theory as it wasn't present at the time.

Finally, fur is the most common type of hair found on other animals. It is similar to hair and its period for phases mentioned above dictate its length.

Two hypotheses for human hairlessness which garner the greatest attention in the annals of paleoanthropology is thermoregulation and parasite removal.

The idea behind thermoregulation is that, with extended periods of time across the jungle wilderness, we lost our hair to promote better thermoregulation - staying cool. This is a wonderful idea, but nature and physics do not bear it out. Studies of both sheared cows and sheep demonstrate that hairlessness is not an effective means for staying cool. In fact, the wool on sheep actually acts to keep the animals cooler. This makes sense when you consider camels and their "hair." These desert creatures are not hairless, yet they've adapted to temperatures that would kill an exposed human in similar circumstances.

Parasite removal is an interesting theory for hairlessness because it supposes that we moved from an area where there were fewer parasites, to one where there was more. Many animals coexist in the same environment and we were no exception, yet we were the only ones who lost their hair? Other primates hunt. Other primates are carnivores. Other primates have had to adapt and shift as the result of a changing environment? Why then were humans the one to shed their coat?

Consider the move from a forested environment where early hominids lived in trees, to one where they lived along the coast. Then consider the hunters now seeking food from an aquatic location. Anyone who has long hair knows the difficulties of maintaining a clean "mane" in that environment. Matting, dirt, and skin conditions (read dandruff and beyond) become problematic.

If you could lose your coat in these circumstances, then an aquatic lifestyle finds advantage with hair loss. Also, hair ceases to be a good thermoregulator when wet. Parasites, which would prefer a damp location, no longer have a home. So, in a sense, the parasite theory, the cooling theory and the coastal theory are all advantageous in a semi-aquatic lifestyle.

Today, zoos across the country are careful when handling primates, not only because of their strength and size, but also with the ease that skin diseases cross from human to primate and back again. Sometimes it is external (lice, fleas, chiggers, etc.), fungal, bacterial or viral. In many cases the primates suffer other ailments which make them susceptible. A compromised primate immune system will trigger other events including skin rashes. Humans are less susceptible because we live in a relatively sterile environment. Water is the best way to purge the infecting maladies, leading again to the Coastal Theory as a viable contender.

The corrolary to loss of hair is increased oil production on the skin as a mediating factor for its exposure. Interestingly, puberty is when the body kicks into overdrive for oil production (more on that later), which would act to assist the hair to fend off moisture and insulate better. This would occur at a time when society demands of early hominids would dictate that the "adult" would start hunting. It is also a time when beard production (on men mostly) and scalp hair reduction begins.

One argument for hairlessness postulated by those touting the Aquatic Ape theory is that other animals who favor water do not have fur. Manatees, whales, elephants, hippos, rhinos, and dolphins are all hairless (note for accuracy sake, we mean that they possess hair, but - like us - it is not a covering matte of fur). This, then is a matter of adaptation. Why does the lion have fur the same color as the dry grasses? Why does the arctic fox have white fur?

Let's consider the zebra. Why would an animal living in that environment have such a wild pattern of light and dark? Most to the predators hunting zebras identify their prey through the visual cue of "edge identification." They look for the outlines of the animal to dietary its type. You'll notice that zebras walk in a staggered line. They collect in groups and never stray far from the pack. Lions, cheetahs, etc. Cannot distinguish one zebra from another and are therefore confused until one animal separates from the rest.

Therefore, if we found hairless to be an advantageous adaptation, we need to look at other related animals who have done the same. Those listed previously are but a few. Of course we could look at the Somalian mole rat who spends his tie exclusively underground.

For animals like the silverback gorilla fur is used as identification. Indeed for many animals the pattern recognition of fur coloration is a distinctive social mark as a means for determining the heirarchy among the pack. Some animals like zebras and tigers have coloration of skin that matches the patterns of hair growth.

We humans have similar patterns in our skin! These stripes called "Blaschko's Lines" wrap around your body no unlike the markings of the fabled characters from the movie *Avatar*. In the 1900's the German dermatologist Alfred Blaschko noticed that skin diseases travelled across the same patterns in different patients. His original sketches illustrated but a simple subset of a rather intricate building of lines as our body matures and skin grows. It is not visible to the naked eye, but can be seen under UV light. Predating our hairlessness, it is quite possible the early hominids carried hair whose colorations matched those found in Blaschko's lines. Today there are animals who can see into the spectrum that we cannot and, therefore, see us completely different than we see ourselves.

We cannot separate the evolution necessary for skin as distinct from the evolution of hairlessness. Where hairlessness reduces the possibility of infection from parasites, it increases our probability of cancer and rashes which often follow Blaschko's lines. Without our ability to produce oil, the skin would not tolerate the conditions found five million years ago.

"Nevertheless, hints have emerged from large-scale comparisons of the sequences of DNA "code letters," or nucleotides, in the entire genomes of different organisms. Comparison of the human and chimp genomes reveals that one of the most significant differences between chimp DNA and our own lies in the genes that code for proteins that control properties of the skin. The human versions of some of those genes encode proteins that help to make our skin particularly waterproof and scuff-resistant— critical properties, given the absence of protective fur.[41]"

The conundrum for hairlessness as an evolutionary explanation makes no sense unless viewed in the context of the Coastal Evolution Theory. Hair loss and hairlessness provides advantage in a semi-aquatic environment. Increased oil production on the skin occurs at a time when the adults would begin a longer aquatic phase of life. When seen as a holistic set of circumstances, hairlessness only makes sense as we wandered from the shoreline out toward the blue.

DON'T SWEAT THE SMALL STUFF

I sweat real sweat and I shake real shakes.
— Elizabeth Taylor

It seems the Bible is full of contradictions, but the story of the empty tomb is one that stands out. Who was first? The women, John, Peter? That depends on which story you read. Was it morning? Was it still night? Were there three women or two? Was there an angel that appeared? All of this depends on which story you give credence. Some pastors have gone to extreme lengths to justify these contradictions. I heard one who said that John was the first but Peter ran faster so he got there before John.

Like those calisthenics performed in the pulpit, the Aquatic Ape Theorists have tried to tie sweat to the evolutionary necessity for an aquatic lifestyle. The scientific crowd has gone to extreme lengths to justify sweating in humans as a natural outgrowth of the savannah lifestyle. Some have gone as far as saying that the naked body was somehow more sexually attractive with a bizarre explanation of "reverse plumage" whose shimmering body was made more wonderful by the glistening of sweat. In the end, whether attractive or not, it performs a vital function for the hairless primate called human - thermal regulation.

But, if controlling body temperature is the point for our sweaty evolutionary path, it is a horrendous move. Many, following the savannah theory, look at the poor hunter human who's only recourse is to outlast the prey, wearing them down with persistence rather than physical superiority.

We all sweat, and especially in the "cultured societies", there is a stigma associated with sweating. Billions are made to cover up body odor. Whether emanating from the armpits, feet, or other darker regions we pride ourselves on our purchased "scent" which ironically is a manufactured bouquet unlike anything that really smells like humans.

The perfume industry is a $38.8 billion dollar industry. Ironically, this industry is mostly self regulated with most studies never published in scientific journals. The International Fragrance Association (IFRA) and their research arm, the Research Institute for Fragrance Materials (RIFM) place few restrictions on what actually goes into perfumes and allows the following:

- Known carcinogens: styrene, pyridine, or benzophenone
- Phthalates, of any kind
- Synthetic musks, including Galaxolide or Tonalide[42]

In fact, our attempt to cover up and prevent sweat may be killing us. The aluminum-based components found in antiperspirants have been tied to specific types of breast cancer. Studies have shown that a higher proportion of cancer cells occur near the armpit in women's breast cancer. Other studies have shown antiperspirants associated with kidney disease. Perfumes have been linked to hormone disruptions, allergic reactions, skin disease, and the deadening of olfactory senses.

So we must ask: Is sweating a product of coastal evolutionary necessity, or a byproduct of hairlessness or other evolutionary adjustment? To answer that question, we must better understand sweating!

Did you know that there are different kinds of sweat? No, I'm not talking about stress sweat vs. workout sweat. There are more than one kind of sweat gland, each with its own purpose. There are, in essence, two types of sweat glands.

The first type of sweat gland, most common to our understanding of sweat is the eccrine gland. These glands cover most of our body. This is yet another human anomaly. In most other animals, there are but a few patches of eccrine glands located on the hairless pads of the feet. For humans, the primary function is thermoregulation. In animals, this provides a protection from hot sidewalks, rocks, etc. But play little to no role in actual cooling. For animals, the role of the eccrine sweat gland is used in combination with its sister gland the sebaceous gland (which produces the oily sweat) to create a barrier to prevent infection from bacteria and other pathogenic organisms. The eccrine gland's sweat mostly water, salt, potassium and other trace elements. One recent study has found that the eccrine gland may also produce stem cells. This is crucial for repair after a laceration of the skin[43]. The eccrine secretion has also helped in providing a medium for friction, thus aiding in grasping and control while in motion.

The second type of sweat gland is the sebaceous sweat gland. When you sweat and it smells, this is the culprit. Let me introduce the sebaceous gland as every teens' worst nightmare. Puberty turns these glands into overworked factories of oil (and acne) producing wellsprings. They are found everywhere on the body with the exception of the palms of the hands and soles of the feet. The oil is called sebum and consists of fatty acids, a waxy substance, trace skin cells and cholesterol. Sebum creates a film that allows the skin to be waterproof.

The sebaceous glands also release vitamin E. This vitamin helps protect the skin from the damaging effects of the sun, reduces the probability of skin diseases, and acts to prevent wrinkling.

The sebaceous glands produce sebum which helps insulate the skin from losing water and therefore actually helps *reduce* sweat production (from the eccrine glands).

Suffering from an overabundance of oily production from the glands may be linked to an overabundance of testosterone. Most of the research that has gone into understanding this gland surrounds its control for healthy skin and less for its origin.

We are not the only animal that sweats. Many animals sweat, though not for thermoregulation. Horses, cows, and hippos are some of the more common animals that are known to sweat profusely.

Primates sweat, but the primary purpose is not for thermoregulation. Because they possess significantly fewer eccrine glands, this is not an effective mechanism for cooling. However, primates with a prehensile tail have eccrine glands along the section where the tail comes into contact with a foreign substance. As with cats and dogs, this produces a substance which provides friction and thus a better grasp of objects.

Thermoregulation, when discussing with anthropologists, always ties hairlessness and sweating with our move to the savannah. The issue with hairlessness is discussed elsewhere in this book, but the evolution "sweat advantage" is problematic. Imagine the hot day when our ancestors were in the plains hunting. Visual focus is key to both eat and not be eaten. Yet, and I'm sure you've experienced this, perspiration gets in the eyes, burns and forces you to wipe it away. This is horrible for human survival and flies in the face of advantage with the Savanna Theory. Also, we as an animal, waste both water and salt during the sweating process more than any other animal on the planet. This requires us to consume more liquid to stay alive. The overall recommendation is to drink eight eight-ounce glasses of fluid a day (which equates to about 2 liters/day), though you can survive on much less. Dehydration works against us in that environment. Consider that with an average workout we lose about .8 liters of water an hour (or about .8 quart). To put that in perspective, it is about the same amount that the standard bicycle water bottle holds. Multiply that by a hot savannah afternoon and you have a very thirsty hominid. Water becomes crucial.

We have almost twice as many sweat glands than our closest primate cousins[44]. Other primates, like the patas monkey, does sweat, though not as profusely as humans. So, under what circumstances do a disproportionate number of sweat glands work under a coastal evolution (we have about 2.5 million glands)? While the scientific community has focused on thermoregulation, the primary benefit in a coastal environment is reduction of biological contamination. The edge of a river bank, particularly where there is still waters, breeds bacteria. The layer of sweat medium creates a protective barrier when immersed in water. When out of the water, the eccrine glands take over and perspiration works like a cleansing agent for the pores and helps remove the contaminants from the skin. The effectiveness of sweating is compounded when you factor hairlessness. The skin is cleaner because hair mattes and collects the substances which would increase the probability for infection.

From an evolutionary standpoint, the sebaceous glands increase production during the period when humans would become fertile. During this period, the "adults" would be primary in foraging for food. The increase in oils works in favor of survival for repetitive immersion in pools of water. As water acts as a solvent to the oils built up, the increased glandular production helps restore the protective layer on the skin that would otherwise dry out and become infected.

Another human trait associated with both eccrine and sebaceous glands is the crisscrossed channels found on human skin which helps channel and distribute the oils across the body more effectively.

As a side note, the palms and feet do not have the glands as previously noted. They do however exhibit one profound human trait - pruning. Anyone in the shower or bath long enough has seen this. It is unique to us. Though often tossed away as some act of physics, it is a *neurological* adaptation. Unlike cats and dogs who *do* have these glands on the bottoms of their paws, we have pruning. Without fingers and feet pruning, our underwater tasks would become incredibly slippery. Yet, pruning provides us the ability to grasp and not slide while immersed. Studies conducted on patients with neurological damage to their hands demonstrate the individuals fingers unable to prune!

But what about that moment in puberty where we begin to grow hair? The locations' areas of abrasion most associated with bipedal locomotion (armpits, groin, and legs) has hair which holds the sebaceous oils. This acts to lubricate those areas and prevents chaffing.

Imagine the savannah scenario as you walk out mid-day into the grassy knoll waiting to attack your prey. You are with a group of others trying to coordinate your activities. However, when you raise your head from the shade, you begin to sweat profusely. Your body is drained of the necessary fluids to help you perform at your peak. You have trouble seeing the others hiding in the scrub because your eyes burn with the salt. Your tongue goes dry, and as the hours wane, you actually stop sweating and feel nauseous. Welcome to dehydration. You become delirious. You are not prey, not predator. It only took a few hours.

Common sense dictates that we would spend the heat of the day in one of two locations: either the shade/cave/ protected area, or at a water source where you have a ready supply of water. We consume more water than other terrestrial mammals. Unlike other animals, we do not pant to help with cooling.

Many of the Aquatic Ape Theorists tie the oils and fat to our ability to stay in the water for an extended period of time. They also talk about how the hair follows patterns for removing water. As a SCUBA diver, I'm also aware that the body loses heat twenty-five times faster in water than on land. At 80 degrees fahrenheit (27C) on land I am quite comfortable because I can generate an aura of heat around my body quicker than the air can sap me of the same heat. However, in water, the same temperature will cause me to shiver should I stay in too long because I cannot generate heat fast enough to keep up with the heat loss. Though this temperature sounds comfortable, after a few hours in water, it becomes quite cool.

That is, again, why the Coastal Theory makes sense. The combination of both land and water provide the warming environment, and supports the reasoning for sweat as both heat release when on land and the oils for skin protection when immersed in water.

THINNED SKIN HUMANS

What's the difference between a taxidermist and a tax collector? The taxidermist takes only your skin
— Mark Twain

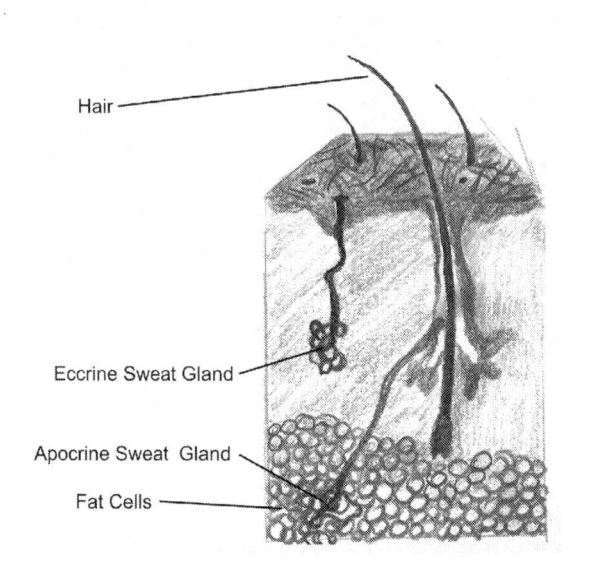

A Body of Water

The hot August sun backed the parking lot as my mom unbuckled the seatbelt of the Plymouth Valiant. Seatbelts were a new thing to me. Our previous car didn't have one and my sister and I would often slide around the back seat when dad took a curve a little too fast.

For some of you readers, you may have considered these the dark ages. There was no McDonalds in my town. There was no pizza place - not yet. We did have the A&W root beer place, and Castle Farms made the best ice-cream known to mankind. Everything but People's drug store was closed on Sundays. There were three television networks and two UHF broadcasting stations that all went off the air sometime around midnight. I'd seen my Saturday morning cartoons and looked forward to the surprise my mom had in store for me.

The line of stores along this strip included Woolco, Peoples Drug store, Mays Hardware, and the A&P grocery store. The parking lot was large and never full. But today it held more than just cars. On occasion, the parking lot would be transformed for a small concert, arts and crafts weekend, or the odd community celebration. That day, so many years ago, in the middle of the tarmac were a couple of semi trucks and a large blue thing with a metal ladder and platform.

We approached the man with cigar in his mouth, paid our entrance fee and proceeded to the above ground pool that had been erected in the middle of the grocery store parking lot. It smelled of fish. Along the far side was an aluminum deck for us smaller kids. I raced up the stairs to see what was splashing water over the edges.

A man in a tee shirt and bathing suit helped me peer over the edge. There, swimming in fast circles was a dolphin. I'd never seen a dolphin before. I don't even recall having read about one in my many children's books I'd inherited from my six other siblings. I had no context to fathom what this large creature was doing in Maryland, so far away from its aquatic home.

But there it was. It did a few tricks. It flew into the air and poked its nose on a ball tethered to a string some ten feet above the pool. I was more afraid than amazed because I thought it might attempt a horizontal leap and end up hurting itself, smashed against a car.

The guy in the swim trunks, sensing my discomfort came over to me and my mom.

"It's okay, boy. It's safe."

He did something in the water. I wasn't sure what.

But the dolphin came over and stuck its head over the pool and gazed at me with a perpetual smile that set me at ease. I smiled back. I liked making friends but didn't quite know what to do next. Should I introduce myself?

The man in the trunks picked me up and I looked over to my mom for reassurance. She was beaming, so I let myself get into the moment. I was the boy, picked out of the crowd for this special event.

"Go ahead, pet him."

I tentatively reached out my hand and felt the rubbery skin, so alien, cold, and soft.

The dolphin shot its head up, splashing me with a little water before it took off, back to its laps around the little pool. I jerked back, laughing and wet.

When I went home, I drew pictures of dolphins and vowed to go to the library and find a book on dolphins. There were none that satisfied my needs, so my dad took me to the "big boy" part of the library. There he found one of the books in the series by Jacques Cousteau. I couldn't read the text, but there were ample pictures to satiate a kid's curiosity.

Looking back on that event, I'm torn between glee and remorse for that dolphin. I never got to know its name. My interest waned as bicycles, UFOs, caterpillars and general kid life supplanted that momentary gift I'd been given. That dolphin was an ambassador for kids like me who otherwise may never have known about these stunning animals. However, he was a social animal forever separated from his own kind, left to entertain and travel from parking lot to parking lot.

I'll never forget petting that strange animal. My mother's gift would eventually have a life-long shift in who I was to become.

She was always there for me. I loved my mom and many considered me a "mama's boy." I'll take that. One of the last times we took her out she was frail and her skin translucent. She fell and she bled. I panicked because my love exceeded my ability to help her. That is the way it seems to get with old people. They look like they are wrapped in Japanese wall paper.

Calling someone "thinned skinned" or "thick skinned" comes from somewhere around the 1600's and is derived from investigating fruit. Either being thin or thick skinned would indicate something wrong with that fruit. So, too, it is with humans.

Two scientists, Adrienne L. Zihlman and Debra R. Bolter, were able to conduct autopsies on a number of bonobos. In their research they compared the anatomy to humans and found that bonobos are "thick skinned" compared to humans. Comparing body mass, the bonobo is composed of 10% to 13% skin - humans are about 6%. They postulated that the thin skin of humans is a result of our hairlessness and ability to sweat[45].

Thin skin is also advantageous for the Coastal Evolution Theory because it provides the increased elasticity to be able to wrinkle. We have evolved the ability for our fingers and toes to "prune." This is an autonomic nervous reaction - not a matter of physics - and is related to the same mechanism that triggers the Mammalian dive reflex. The increased surface area allows for more tension when immersed making grasping easier. This "glabrous" skin isn't random, though it may appear so. In fact, it has a peak in the middle of the digit and wrinkles away, providing channels for the water to move when pressure is applied to the digit's surface. Our ability to grasp underwater would be crucial for hunting for eggs, crabs, shellfish or other food sources found along coastal boundaries.

Thinner skin, however, lends itself to an increase in the probability of damage via UV radiation. Hence, there is a need to protect the skin. Scientific studies have shown that exposed skin results in an increase in the production of sweat from the sebaceous glands[46]. The sebus produced creates a protective barrier from the sun and creating the anti-bacterial layer mentioned previously. Modernity, in its efforts to remove that layer from the skin less we look "oily" may be the actual thing that is making us less healthy. We compensate by applying UV protection to the skin via the plethora of commercial sun blocks commercially available.

There has been an undercurrent in the media saying that sunscreens may actually *cause* cancer. The scientific evidence is still out on the subject. Of particular interest is *oxybenzone*, a chemical that is known for its power as a hormone disruptor. Add to this other chemical agents working as a topical agent with the sun's power to energize and transform the stew applied to the skin of sunning tourists around the world. It is clear that oxybenzone and octinoxate are hazardous to fish and coral reefs, so much so that Hawaii prohibits the sunscreens that are killing the the fragile paradise that are the islands.

Underneath the layer of the skin is fat. In humans, the layer of fat is markedly greater than other primates. Additionally, as infants, we are born with a greater layer of fat - aka - pudgy babies. In fact, we are one of the fattest born animal on the planet. Our babies are born with about 15% fat and that grows to a whopping 25% by four months. But is baby fat the same as adult fat?

Many of the Aquatic Ape Theorists pose that the fat of the baby is somehow related to the fat found in dolphins. It is used for buoyancy. They cite the way that babies naturally hold their breath when submerged and are positively buoyant in water. The "Aquatots" programs tout that babies are fine when submerged because they have a natural reflex to hold their breath. This is a dangerous assumption and only partially true. Yes, toddlers do have a natural reaction but many can do this for only a short period of time. In fact, babies swallow water and can suffer from water intoxication resulting in seizures and death. Hyponatremia[47], even if non-lethal can create a fear of water in later years. So, if babies aren't born for the water, does the amount of fat dismiss the Coastal Theory as well?

We actually have two kinds of fat: white fat (your basic McDonalds diet style fat) and brown fat also known as brown adipose tissue, or BAT. BAT is important for babies because it actually generates heat. BAT converts the white fat into heat, and can be used as heat reserves for situations when they need nourishment and there isn't a ready supply (sometimes it takes a while for mothers to develop their milk). This energy is also important as human babies have such large brains. The brain consumes half of the infant energy store. So, in conjunction with generating heat, the fatty tissue supplies energy to the brain much like a battery[48]. Additionally, recent studies have shown that active BAT in children promote more muscle and bone development[49].

So the question arises, did fat babies come as a result of the need for energy stores with a change of environment and an alteration in dietary needs? Did fat babies arise because of the incredible energy required to feed the brain? Did the change in our musculature and frame dictate an increase in the need for BAT? Scant evidence exists to help answer these questions. Also, the timeline for the development of a large percentage of fat in infants is still up for debate.

However, if one wants to argue fat stores for heat, then they also have to argue hairlessness for cooling. The contradiction seems less tenable in a Savannah Theory than a Coastal Theory where thermoregulation is a lesser concern for hair loss. But even that case is weak for the Coastal Theory because scientists have not confirmed that the fatty buildup correlates to bipedalism, and may have occurred at a much later date. Indeed, the fat ratio may be a result of a much later migration to cooler climates.

Though the evidence is still out on fat, skin is definitely geared toward the watery world. Soaking in sea water improves your skin. It releases bad toxins while absorbing good ones. Thalassotherapy (Greek: thalassa = sea) uses sea water, mud, kelp, and seaweed to improve the skin. Scant scientific evidence exists, though there are references to bromine, iodine, zinc, manganese, calcium, sulphur and iron being absorbed as a result of immersion. The treatments tout everything from skin restoration to reduction in cellulite. Like any therapy, it is always good to do your research to find what specific benefits can be had by the therapy. It is also wise to conduct a thorough background check on the individuals and facilities conducting the therapy. Though I've not paid some of the exorbitant fees found at some institutions, a good ol' dip in the Atlantic or Pacific works just fine for me. The effects and results seem to be the same.

As new divers, my wife and I headed to Grand Cayman to dive. It was touted as a mecca and when we arrived we were not disappointed. The dives held a theater of life which performed regularly on our underwater trips. At night we dined under the stars. The romantic spell was cast. During one of those wondrous meals, my wife commented on how our skin felt rejuvenated by the dives. I assumed the sea would dry out our skin, but instead, improved our complexion. I found myself changed by the immersion in the ocean. Beyond the psychological effects of diving and vacation, the physical effects of improved skin are readily apparent.

It is important to note that when performing a series of repeated dives, specifically when "working" as a divemaster or instructor, the extended time in the ocean can make your skin more susceptible to cuts. With pruning of the skin and extended underwater time, I receive more cuts from incidental brushes with surfaces more so than on land. That may be attributed to the fact that I'm in contact more surfaces that may pose a threat, or that I'm conducting operations which present the probability of cuts and bruises (or perhaps one could chalk it up to an old man's clumsiness) but I have far more scars from underwater encounters than at the surface. However, in 2006 David Blaine, a popular magician whose unique take on his craft includes a guru style addiction to extremes once placed himself in water for days. His magic trick failed (he was unable to break a breath holding record), but science took advantage of his time immersed. He suffered liver damage and his skin started peeling away, losing its cohesion after being wet for days on end. This doesn't negate any of the theories, but does fly in the face of some of the edge "mermaid" theories. We are definitely a terrestrial being, who may enjoy time in the water, but is not returning (from an evolutionary standpoint) to our fish ancestry any time soon. We are planted firmly on terra-firma with only a foot in the ocean. But, if we are to fully appreciate both skin and hairlessness as a necessity or advantage in the Coastal Theory, we have to address not only the water and its effect on the skin, but also the effect of the sun.

UV radiation also alters the skins ability to absorb moisture and self-heal. Though it has been known that UV radiation is important with skin exposure for the production of vitamin D, its harmful effects are well noted in regards to premature aging of the skin as well as skin cancer. Keratin strands and the interconnected channels found in the skin help channel water away, while also absorbing water as noted in the change in skin when someone takes a bath. Skin damage will affect the skin's ability to absorb the water and effectively channel it away from the body[50].

Today, Rainbow Reef Dive Center in Key Largo, Florida provides free sunscreen for all of its divers. It is a reef-safe variety that, in the coming years, will be the only available option in Florida. This progressive move follows what is already happening in Hawaii and other tourist sites around the world. Many of the sunscreens are toxic to reef life, so take care to read the ingredients or, better yet, purchase the sunscreen when you arrive at your destination, making sure you select a reef-safe variety. And, if you are looking for recognition from your purchase, take time to SCUBA or snorkel and look at the abundant life you are preserving.

FACIAL RECOGNITION

For he, who as gained control over his breath, shall also gain control over the activities of the mind. The reverse is also true. For he, whose mind is in control, also controls the breath. The mind masters the senses, and the breath masters the mind."
— Hatha Yoga Pradipika

Say the word primate and a fairly static image of a chimpanzee or gorilla comes to mind. But this is far from the norm. Indeed, many primates look surprising different.

Madagascar follows with other large islands like New Guinea in its odd fellows of primate evolution. At some point in the distant past when Madagascar separated from the mainland Africa it took with it a host of animals destined to follow some divergent path of evolution. That is the way of many islands. The isolation forces a split in the family tree, often with surprising results.

Show any child a picture of a Madagascar Lemur and ask them to describe another animal that looks similar and you might get squirrel or raccoon. Yet they are a primate. They are a distant branch of some little changed ancestor spanning millions of years to our primate trunk line. It is not hard to imagine these wide-eyed, long snout animals scurrying past the roaming dinosaurs, unaware that their lineage would, one day, dominate the planetary landscape.

Their asian cousins, the tarsiers, have a face and small ears that look more bat-like than chimp-like. Pottos, galagos, loris, and the wacky looking aye-ayes are but some of the smaller primates we may not recognize in our primate family reunion. Marmosets, also a primate, has both a smaller snout and eyes. Barely the size of a human hand when fully grown, these cute animals display

more chimp-like facial features. Their hands look like larger primates with the exception of claws rather than fingernails.

Larger primates like the langur and proboscis monkeys exhibit are incredible to watch because, like the larger apes, they exhibit "human-like" facial expressions. As we follow the lineage from the smaller, more squirrel-like animals to the larger apes we see a greater and greater similarity to the image we see when we look in the mirror. Looking at different primates allows us to see a line of evolution from the earliest "looking" mammals that eventually became us. Indeed, a controversial study by Jeffrey H. Schwartz, professor of anthropology in Pitt's School of Arts and Sciences and president of the World Academy of Art and Science, and John Grehan, director of science at the Buffalo Museum, argues that humans have more physically in common with the orangutan rather than the chimpanzee. Because we have not properly or so incompletely mapped the human genome with other primates, the genetic comparison is far from conclusive. They made comparisons to the thickly enameled teeth, mammary glands, hairlines among other traits we have in common[51]. Because the evidence is still out and genetic evidence supports the chimp-human connection, many consider the theory as "wacky." It is clear that our genetic heritage has more in common with the bonobo than the orangutan, but it is unusual that we share so many similarities in our construction so that it cannot be ignored and warrants further research.

The human head and facial features are a radical redesign of the architecture of any other primates. We have gone through an amazing rebuilding of the head which gave us the ability for complex vocalizations, updates to breathing, changes in our brow and a host of other new designs due in large part to our vertical stance. This called for a redesign in how we operate everything from hunting to communicating.

Famed comedian, movie star, and talk show host Whoopi Goldberg shares her home with a dog named "Bear." She feeds and walks the dog like any other canine owner. However, bear is both deaf and blind.

"For me, it is a question of who rescued who because I didn't know how much I needed this dog until he showed up," said Goldberg.

Dogs are olfactory animals. They can smell things we cannot. They can also smell in *ways* we cannot. They tell time based on scent, and they have an acute stereoscopic sense of smell. So "bear," though lacking in senses critical to us, are not as encumbered should we face the same fate. In fact many animals rely on scent as their primary locator for food. That is not to say that humans have a poor sense of smell. In fact, we can discern smells as well as many other animals on the planet, but something changed along our evolutionary path that rendered our olfactory hunting wanting.

Though we may be able to detect smells, like "Bear," we do not rely on our sense of smell. A study from UC Berkley confirmed that we do have the ability to track scents to some degree. Placing a tracking device on hunting dogs, the researchers were able to see the path the canines took when following the trail of a pheasant. As expected, the dogs would veer from the birds trajectory and make their way back. This formed a zig-zag pattern.

UC Berkley students did the same thing with a number of volunteers. If you happened to be there that day, you'd see a series of students smelling the grass as they tried to locate chocolate scent laid down by the researchers. Until then, scientists assumed that our poor sense of smell would send the students in random directions. However, the students didn't do bad. No, they weren't as acute as tracking dogs, but they were able to follow the path. In fact, the track that the students took resembled that of the dogs tracking the pheasant. The "waiver" or amplitude off the track was larger than a dog, but that was to be expected. As students repeated the test, they performed better. The researchers noticed that humans turn their head from side-to-side much like other animals tracking a scent. In doing so the small receptors at the base of the nose can detect which "side" the smell is coming from, thus mimicking the better developed sense of smell of canines stereoscopic olfactory system.

The result is that our scent-based sensory acuity has not atrophied as some aquatic ape theorists suggested. We can detect the relative location of a scent like most other animals. Yet, our evolutionary path has relegated our sense of smell to only a passive part of our hunting instinct. If not by biological design, then why did we choose to opt for other senses to become the primary detectors for food?

Some would say that the answer is a plain as the nose on your face. Indeed the structure of our nose is yet another anomaly in our primate construction.

Our nose holes face downward.

The most current theory is that the down-turned nose performs two primary functions. First, the downturned nose allows greater capacity to smell food entering the mouth. As a kid, you may remembering holding your nose. If your parents made you eat something you didn't want, then you may have pinching your nose while consuming whatever vial substance forced upon you. For me, I still have an aversion to ketchup and pickles. Pick your poison.

The chemosensory combination of smell and taste is important in humans because we thrive on the variety of food. As Omnivores our combined sense of smell and taste work in tandem to prevent us from consuming poisons. Other animals with a less varied palette have a different configuration for taste. Cats cannot taste sweet. Many Aquatic Ape supporters compare the nose shape in relation to other aquatic mammals. This is a false equivalency because the roots of our evolution do not compare to sea lions, etc. In fact, many of coastal animals do not have as good a sense of smell or taste because their diet of fish is usually swallowed whole[52]. Therefore, the position and shape of the nose would have no bearing on the outcome of the diet. Our culinary discourse relies on smelling as part of tasting.

The second reason for the downturned nose is thermal regulation. The modern nose has a thinner set of open nasal cavities. The path that the air now takes is far more circuitous than in ages past. It doesn't take much scientific investigation to note that in hotter, tropical climates the natives have broader wider openings (where the air needn't be warmed) versus colder climates where the natives have a more pronounced proboscis with a thinner width. This makes sense, but there is one problem.

The projection of the nasal re-architecture started well in advance of the neanderthal and any travels northward.

So this blows the second theory for our olfactory redesign. If this predates any migration to colder climates, then the re-architecture must have served some other purpose.

As a consequence of the new facial architecture, the downward pitch of the nostrils also works in favor of someone entering the water vertically. The air is trapped in the nasal cavity preventing water from going up the nose. Most humans can hold their breath and not worry about water entering their nasal cavity without the need to pinch their nose. We can turn our heads, look down, move forward and generally scavenge the watery floor without choking on an incoming flood of water into the lungs.

A Body of Water

During the evolutionary process, our olfactory system in the brain took an interesting turn as well. In a visual and auditory world of humans, the olfactory system moved to the base of the brain and became enlarged. So, the question arises, why did the olfactory system evolve into a larger mass in the brain if our sense of smell as a primary means of survival deteriorated?

"Olfactory information thus projects to regions critical for mating, emotions, and fear (amygdala) as well as for motivation, high-level cognitive and emotional processes (orbital prefrontal cortex). It, thus, serves a role in central nervous system function above and beyond smell[53]..."

As a kid, I lived in the attic of our house on Elm Street in Frederick, Maryland. There was a smell associated with the house which was unique. It wasn't mold, smoke, or anything one might distinguish as some identifiable attribute. It smelled like *home*. It was a smell totally different than my grandmother's house, or my neighbors. I knew what they smelled like also.

It is nostalgic moments when I reflect back on my boyhood. I can still see my mother in her housecoat, making "sticky buns" as the house filled with the smell envious of any bakery.

I would still be in bed in the morning, long before the sun came up as the coffee percolator started singing and the smell of hickory coffee filled my room.

I'm sure you have similar memory tied to smells. Smell seems to take on a more important role in childhood and is lost as we mature. The sense of smell retains a visceral, primitive GPS. We tie smells to location, emotion and association of memory. But the basis of our existence, the essence of who we are as human is, in large part a visual record.

As the human brain developed (largely due to our now erect posture), other areas of cognitive functions evolved and adapted to adjust, This would include such things as language, motor balancing, and a sense of direction[54]. But with the increased size of our olfactory bulb, we relinquished many of the active genes used for discernment of smell.

"... Why have so many of our odor genes been knocked out? Yoav Gilad and his colleagues answered this question by comparing genes among different primates. He found that primates that develop color vision tend to to have large numbers of knocked-out smell genes. The conclusion is clear. We humans are part of a lineage that has traded smell for sight[55]."

As coastal creatures, whose diet would consist of blooms from reeds, crabs, shellfish, aquatic eggs and other gifts from the shallows doesn't require scent to hunt it out. A semi-aquatic environment depends more on visual, than olfactory hunting. Indeed, with the change in the architecture of the face, the nose played another role advantageous to wading into water. Humans have four rather large sinus chambers (which is why we suffer so when we catch a cold). These large chambers store air. The sinuses can be used to warm and humidify on inhalation. However, it also makes the head more buoyant. This is important if a coastal creature wades too far and - sorry for the pun - gets in over their head. Yet, the scientific world sees that the evolution of the paranasal cavities as not related to the a semi-aquatic lifestyle.

"If paranasal sinuses ever functioned to keep hominin heads above water, they did so without any demonstrable evolutionary change[56]."

The question then arises whether this was a matter of directed evolution or a byproduct of bipedalism? Is the overhauled olfactory system a result of our need to wade out ever farther into the deep to find food that may have become scarce?

We can only speculate that if our environment changed, we adapted to fit the circumstances. As we adapted to become coastal hunters we went beyond those first timid steps from the muddy shore. Our success was our own undoing. With the change of environment we entered a period of time where predators didn't see us as prey, not unlike the lion fish that have invaded the Florida coast line. Sharks and other apex predators don't look to these odd animals as food (in fact SCUBA divers have started feeding lion fish to the sharks in an effort to teach them that they are viable prey). So we had a grace period where we could easily escape our land-born predators by entering the water while finding free rein in the wetter part of our ancestral home. We flourished and the population grew.

With the increased numbers our food consumption increased. With fewer resources along the shallows we were forced to move ever outward into the water. Those with more skills at breath holding succeeded. Those whose nasal architecture favored the deep survived. They learned to swim.

Have you ever accidentally gotten water up your nose? This painful activity usually results in a quick surface, coughing, gagging and snorting. But some free divers have learned to accept the parasympathetic reflex associated with water entering the sinuses and ear cavities. It is training that starts with a neti-pot activity.

The neti-pot looks not unlike the genie in a lamp. Using distilled and purified water[57], free divers mix this with a saline and pour in one nostril with the head turned. Holding back the gag reflex as the water courses through the sinus cavities and exit through the other nasal opening, they allow the water to channel through the back of the throat. Once the free diver gets a handle on this exercise through repeated attempts, they "feel the burn" and relax the jaw muscles, moving it slightly from side to side. This allows the water to fill the ear passages. With repeated practicing the free diver can inhale the water to the point that it fills the crucial portions of the nose and sinus cavities that would otherwise need to be equalized during a dive[58]. Because water is liquid and therefore does not compress with increased pressure, the sinuses and ear canals are not at risk of barotrauma.

Freedivers a.k.a. Apnea divers are the extreme example being able to breath-hold and plunge to depths once thought unimaginable. But if we look at our ancestral coastal ancestors, they would have been challenged by forces, whether flooding, current, deep water, and underwater obstacles that may have led them to *swim*. If the water was only a few feet deep, they would still have to breath hold to reach down and grab the food source from the murky bottom.

So why did *walking* positively affect our now buoyant head? If our olfactory architecture was not the result of our new hunting practices, how did we get to where we are now?

With our head now planted atop our body, the mechanics of walking requires an entire new method for locomotion. I was in an airport and, while idling before a flight, I "people watched." The airport is a great place to observe human locomotion in action. One extreme case passed my gaze. She was an attractive blond professional who wore a tight knee high skirt. She was in a little bit of a hurry. In one hand was a Starbucks coffee without a lid. She was talking on her cell phone. So both hands were occupied and thus unable to swing with their natural motion. She had on high heels. This, plus the skirt prevented her from extending her stride to meet whatever arbitrary deadline she was trying to meet. It was the most artificial looking walking style I'd ever seen. The woman had literally strapped herself into a position of disability with her choices as she didn't so much walk as slide her feet a few inches at a time. I could only imagine the stress on her back as she tried to not spill the elixir of coffee she purchased. Her arms were of no help in her locomotion. Her head, the final bastion of balance was cocked to an odd angle and seemed locked in invisible stocks, unable to assist in her propulsion.

As she shuffled along a small kid half walked, half danced around her until his parent called her back to them. This kid was wildly swinging his arms as he took deep steps, and twirled to an unknown rhythm inside his head.

These were a study of opposites, with the child faring much better in the end.

I know the suffering of back pain. I try to run at least three times a week. I am rather tall and the majority of my height is centered in my torso. So, with my great height, I retain a higher center of gravity than others my size. This means that my pendulum is as pronounced as my vertical peers. After my workout, I'll spend another fifteen minutes doing stretching exercises. I hope, in one particular twisting motion, to hear my back crack. Otherwise I run the risk of a day with the potential back pain.

Human running is unique, as we are a bipedal quirk of nature. When running, a dog keeps its head still along the thorax. Because it is a quadraped, the muscles are thick along the neck as it juts its head forward. We slow running humans also need to keep our head stable, but because of the way we run we have had to evolve an incredible reorganization of biomechanics. This is one reason we now lack the power of our primate cousins. We swing our arms as opposing pendulums to each other which weigh about the same as our head in an effort to stabilize this operation. Because of the loss of musculature in the neck and because of the restructuring of muscles for running, we lost more advantage than we gained. Our head, so much heavier for our size than any other animal is situated delicately on our neck which is thin and weak. So where does this work in our favor?

If you have ever seen a cargo carrier, or military style submarine, you will notice a bulbous creation in front of the bow. This is known as a *displacement hull*. The water is displaced, allowing the craft to move more efficiently through the water. This works at slower speeds as the water moves around the object in motion. It creates a space for the larger vessel to move without having to *push* the water out of the way.

The human head seems large and heavy out of the water but when immersed in fluid the head is neutrally buoyant and acts perfectly as a displacement hull. I'm not postulating that this is the reason that the head shape and size was the result of a necessity for this sort of adaptation, but it is a fortuitous byproduct of the change. When we swim our head plays a crucial role in moving water out of the way to benefit the rest of the body.

The realignment of the spine facilitated a change in the facial structure. Our jaw receded as a result of the realignment of our neck. The muscles associated with the jaw weakened, creating a look far different from bonobos. Researchers say that the loss of the muscle structure in the jaw was one of the primary reasons that we have the size brain cavity that we possess today. The loss of *Facial Prognathism* or facial flattening not only increased our nasal cavities which was beneficial for swimming, but also resulted in the lowering or our larynx. The descended larynx made human vocalization easier, though plays lesser of a role in the production of sound as previously thought. Indeed, many other animals share the ability to create complex sounds. One distinguishing feature that separates us from primates isn't so much the physiological differences, but rather our ability to hone *"vocalization control*[59]*."*

"The development of language occurred with the evolution of the prefrontal cortex. This part of the brain houses what is known as *Broca's area* specifically geared toward the formation and comprehension of language.

French neurologist Pierre Paul Broca described a patient unable to speak who had no motor impairments to account for the inability. A postmortem examination showed a lesion in a large area towards the lower middle of his left frontal lobe particularly important in language formulation. This is now known as Broca's area.

The clinical symptom of being unable to speak despite having the motor skills is known as expressive aphasia, or Broca's aphasia[60]."

The assimilation of words uses the Wernicke's area and the angular gyrus, in tandem with auditory and optical cognitive and motor processes. So, unlike Broca's aphasia, Such development of the brain resulted in the increase of the total brain size. Wernicke's aphasia presents when an individual can speak, but not comprehend language. Disruption of cognitive functions regarding linguistics under the individual's ability to communicate.

The requirements for human language, therefore required minimal physiological change from a motor function standpoint, but a much larger alteration in brain function and development. The change from a minimal, mostly "sign language" approach to communication, to a sophisticated auditory language could have been sparked by a coastal lifestyle.

Though this is speculation, it is easy to see that one loses the ability to physically communicate if partially immersed in water. With the hands and feet otherwise preoccupied, the only alternatives reside above the shoulders. In the savannah, vocalizations would work against the hunter, alerting both prey and predator alike. However, language becomes critical for locating each other, or offering direction while partially submerged.

Humans, under these circumstances would have to transfer a physical lexicon to an auditory lexicon. From translating symbols, to a primitive syntax would require development of the brain consistent with what is found in paleontological evidence. Indeed, the Broca's area is well developed in chimpanzees and shows activities in these primates.

Gwen Bell Darby was in a bar carrying on a conversation with a friend of hers. It seems they both had a little to drink and, with the alcohol removing some of the social filters, were making comments about all of the people in the bar. There was the "hot" guy, the "loser", the "slut" and a host of others that required their review. They didn't hide who they were talking about. No one seemed to mind, in fact, many were staring at them while they commented on the clothes or hair. The two carrying on this conversation were not sitting together. In fact, they weren't speaking at all. Gwen, an expert in primate linguistics, was gesturing using the language of the Bonobo. With a series of facial and hand gestures, they were able to carry on a detailed conversation. The Bonobo sign language is a developed and detailed series of gestures that was known to the two scientists, but not to the victims in the bar. Body language tells stories. On many occasions I've been able to read chapters from an individual by the way they carry themselves. My wife knows when I've had a bad day within seconds of crossing the door. Sit at a park, mall, library, or any public area and you can read the body language of an individual and, with some accuracy, identify some general truths about that stranger. Whether we know it or not, we are attuned to body language as well as the spoken word. And, the tools by which we use to interpret body language reside in some of the more ancient areas of the brain.

We are a land of many languages, each still evolving and mutating, as society changes, technologies arise, and the lingo of a new set of professions emerge. We sing, shout, whisper and change our voice to mimic others. Of course language is an abstract construct and leaves no obvious trail of bones. One argument for the development of language came from the facial architecture changes that occurred as a result of bipedalism.

Today our verbal repertoire allows for communicating abstract ideas, emotions, inferences and abstract concepts. Our ability to mis-interpret the same is daunting. We've developed a variety of ways of communicating both verbal and visual languages and new ways to do so are being realized as technology changes the way we connect.

We no longer resemble our lemur-like family line. Consider those who endured the cold ice-age in Asia. They bear the mark of the epicanthic fold over the eyelids which represent the look of an "oriental" human. Those who baked in the African Savannah, when the verdant forests turned into a desert wasteland bear the dark skin of "African American." Large noses, balding, brown eyes, and other facial and bodily anomalies are born from some alteration in the environment or some force that required the human design to rework itself to become better adapted. Likewise our face changed as a result of the years we paced along the muddy edges and estuaries that eventually turned to sand. We proclaimed our independence of the treetops only to find the harshness the planet earth denying the very thing that made us who we are. We were left to roam northward and outward to find whatever pools of water would bring relief. Our habits had changed. But more than that, our bodies changed, forever denying our return to the shady limbs high overhead. And, when we did find those forests where our relatives jumped from branch-to-branch, we looked up and saw something different - something not-us.

BLUE GENES

The land may vary more;
  But wherever the truth may be-
  The water comes ashore,
  And the people look at the sea.
              -- Robert Frost

In 1970, British mathematician John Conway created a computer simulation which demonstrated evolution, known as "the game of life" using only a few simple rules. Consider a pixel (a square in space like that of a chessboard) - a single square is either "occupied" or "vacant":

- For each cell that was "occupied"
  - Each pixel with one or no neighbor dies, as if by solitude
  - Each cell with four or more neighbors dies, as if by overpopulation
  - Each cell with two or three neighbors survives.
- For each cell that was "vacant"
  - Each cell with three neighbors becomes populated

Experts in the "game" have developed scenarios where populations explode, tribes migrate and form knew populations. Some collections of "pixels" fight and retreat. The gods of the game can add or remove pixels at will which may radically alter the outcome of the "life." It is a way of looking at a one-dimensional simulation for adaptation and evolution. The basic principle of the game has evolved and been incorporated into games such as SimCity™ and Halo™. Anyone involved with simulator game programming knows how applying these types of rules can dramatically alter the way the game behaves. Small changes early-on can have increasingly greater effect down the road. The placement of a single school, or the choice during a conversation may come back much later in the game resulting in an increase in the overall property value, or gaining an ally in warfare.

We can look at DNA chains and our most rudimentary chemistry and see how these kinds of rules play in our world. The seemingly simple rearrangement of proteins can have dramatic effects. Move one and you have blue eyes. Rearrange a few and you have three hands, twelve eyes, or wings. Science fiction writers have theorized that once genetic manipulation becomes the norm, gaining another appendage will be not unlike getting a tattoo.

The assumption is that we possess static chains which get copied (at least in part) from one individual to their offspring. This, however, is not the case. We have traveling genes, "selfish" genes, paternal genes (genes that knows when they are transmitted from a guy), maternal genes, and genes that apparently do absolutely nothing. We are the consummation of our genetic soup, but that soup is constantly being cooked, added to and sampled. What we do to our bodies during the course of our life, like smoking, exercising, eating fish or watching too much television all has relevance to what happens in our genetic code.

Generating a human is, in itself an extraordinary event. The steps that are required for us to eventually end up as a suckling infant seems impossible. On becoming who we are, we transition through the entire microcosm of our human evolution. Beyond memory, in the darkness of the womb we played out the drama of the millions of years it took to become "us". We began as simple cells, dividing and replicating. In a matter of four to five days we became a group of somewhere between fifty to one-hundred and fifty similar cells grouped together. This group contains the all important and controversial embryonic stem cells. Stem cells are important because they are the precursor of things to come. Though identical, they contain the library for eyes, ears, heart, blood and all of the varied components that make up who we are to become. Science throws up its hands understanding those transcription factors (the mechanisms by which proteins begin to change from one kind of cell to another).

After about two weeks the differentiation occurs. Blood, kidney and nerve cells develop. The basic building block "organism" - the root word *organ* indicating that now we have become a team of cells with specific tasks that represent the whole. Interestingly, as fully adult, we rarely think of ourselves in the plural. We identify as "one" rather than a conglomerate of interdependent working parts, independent and cohesive in form and function. Yet, at two weeks, we divide into a series of units that forms the "us" in "I".

During the early part of the first trimester the brain and spinal cord begin to develop. With the development of the eyes, human embryos develop gill slits and a tail. We are fish. In fact, it is difficult for the most educated physicians to tell the difference between a human and shark zygote at this point in our creation. We have entered our Cambrian development period. In biological time, it is just a month or so, but looking at the planetary calendar it was about 550 million years ago and lasted millions of years.

The simple creature that would look more at home in a fishbowl than in a playroom has a genetic surprise for a few infants. Some births have resulted in the infant with the tail intact. Often this is removed soon after the child is born. The gill slits, too, have been present at some births.

"Rarely, children are born with an acute vestige of an ancient gill cartilage... In these instances, my surgical colleagues are operating on an inner fish that unfortunately has come back to bite us[61]."

The area in our inner ear which allows us to hear sound is the cochlea, a wound nautilus of tissue containing a fluid called *Perilymph* a liquid which helps translate the waves from the ear drum to small hair-like fibers called *Stereocilla*. The perilymph has the same specific gravity and is quite similar to sea-water. The stereocilla wave in the ebb and flow in its watery canal and the results are sent through the Cochlear nerve ending up in the brain as the sound of a laughing child or Led Zepplin.

The engineering of a gill is remarkably similar. Unlike the human ear, there is no closed-off canal. The sea water flows through the gill slits and moves through the gill filaments. These filaments are much larger than the stereocilla but the architecture is similar. The gill filaments act like our lung's aveoli moving oxygen poor blood through the small fibers as the water travels through the small filaments.

Each step in the micro-evolution of human development mimics an earlier version of our evolutionary plan. From fish, to lizard, to human we progress along our ancient chain until we emerge from our watery home as a new-born child. But how did the genetic instruction manual know how to assemble us as it did? And, do we still retain the rosetta stone to be able to translate what we are in some meaningful way?

That library that is contained inside our stem cells are known as the (Homeobox) Hox genes. These genes are responsible for determining our body plan. Whether you are a fungi or a wombat the distinct library for who you are resides within the pages of the Hox gene. By comparing Hox genes across the kingdom of life, we can map and compare similarities and differences. Far from the paleontological science of digging in the dirt for clues to who we are, we now have the technology to mine our own DNA for answers to our watery past.

A two-pronged approach to genetics has helped dispel some of the evolutionary myths regarding our evolution. It has lent evidence to our coastal theory of evolution as we mine both the past - through paleogenetics, and ourselves - through comparative genetics.

Paleogenetics is a straightforward operation, though like its paleoanthropology counterpart is limited to the scant findings available for research. Extracting DNA from ancient bones is an art form and often yields in incomplete strands. Yet, what it is able to uncover is invaluable. We can see strands that we carry unaltered in our present form and see those attributes which are no lost to time. For those missing pieces we can identify what changed (the lack of a brow ridge or a body covered in a matte of hair) and glean what was required for us to evolve from one environment to another.

Comparative genetics takes what we know about us, our DNA library, and compares it against a catalogue of other species. This library is rich and allows, not only for a comparison of how we are different from the other creatures on the planet, it provides a way to detect which specific "switch" has either been turned on - or turned off, to make us who we are.

Let's begin with a baseline. Genetically speaking we are a primate. There is no surprise there. It is common knowledge that we are more closely tied to our genetic ancestor the chimpanzee (*Pan troglodytes*). But there is a cousin, the bonobo (*Pan paniscus*) to whom we are nearly siblings. The bonobo is an endangered species found in the area of the Congo Basin in the Democratic Republic of the Congo located in Central Africa. Because their similarity to the chimpanzee they weren't even considered a distinct species until the 1920's. Using comparative anatomy a scientific study concluded that the bonobo is closer in anatomical structure to humans than the chimpanzee[62].

Studies have indicated that the genetic drift that separates us from these animals is about 1.2%. Calculations comparing the chimpanzee to the gorilla shows a genetic drift about 1.6%[63]. Amazingly then, the chimpanzee is closer to a human than a gorilla! Using genetic ancestry, we can trace the common ancestor of our bonobo, chimpanzee family reunion somewhere about five million years ago. This corresponds with the evidence of our bipedal nature. Genetics also demonstrates a large period of inter-breeding between the bonobo and the chimpanzee making the genetic distinction that much more difficult. If we follow the lineage of our bonobo brethren geographically, then we are looking at our ancestry along the Congo River.

The bonobo are highly complex social animals. They live a primarily egalitarian lifestyle (with the matriarch more often in charge), though families tend to congregate together. Males are more likely to interact across family lines than females. They do not stay in large groups and have a rather diverse diet. They have been known to alter their diet, injecting medicinal plants or animals to quell a particular malady. bonobos also develop cultural differences within their respective parties. Grooming practices change from tribe to tribe and the language they speak may differ. They are the most outspoken of our chimp cousins, articulating a complex array of vocal signals. It is widely known that bonobos, especially during the wet season, will wade into shallow water in search of food[64]. During this time, they exhibit a bipedal stance, though they would rarely submerse beyond waist-high.

Given our genetic similarity to the bonobo and the chimpanzee what are those elements that help define "human?" If our genetic drift is less than two percent, what are those minuscule variations that render you versus your bonobo family member? Yet that is only part of the question. Comparative genetics is relatively easy. The more difficult question to ask is, how did they come into being? What caused the genetic drift to occur? What benefit did we get as a result of our rewriting of the genetic script?

The human genome project in its ambition to completely map our "humanness" is reticent to claim that they've identified about nine percent of the gene library as "active." This is the territory of strands that make us who we are. The rest according to scientists, called non-coding DNA, had originally been thought of as "junk" DNA. This however is not the case. These specialized strands act as guardians for turning on and off specific genes. The junk DNA are the transitory members of our protein soup carrying replication of itself which may transpose pieces of DNA (RNA) and rewrite the story from parent to child during cell replication. This is particularly true for HOX genes.

The "survival of the fittest" in the Darwinian model is now but a subset contained with "mutation of the DNA" as part of the evolutionary strategy. This is a harder pill to swallow because the idea of random mutation plays against divine plan as well as working against the narcissistic view of man at the top of the evolutionary pyramid as a result of millions of years of fighting for the title of "the fittest." Indeed, we may be witnessing the possibility of junk DNA to switch with relevant DNA when a change is necessary for survival. The end result is not a Darwinian progression over a period of time, but rather a rapid shift based on external forces requiring change.

Early genetics proposed that the "message in a bottle" carried to us from our parents was cast at the point of conception, never to be altered. Scientists thought that only atomic level radiation could scramble the genetic alphabet and, more often than not, that mutation would quickly be dispensed with as a foreign agent that wouldn't survive, much like bacteria and the immune system.

But that simply is not the case.

Sometimes there is the genetic hiccup. The molecular strand known as a *nucleotide* consists of a distinct set of molecules. They are:

- Adenine
- Cytosine
- Guanine
- Thymine

Gratefully scientists have shortened these to A-C-G-T. It is the combination of these nucleotides that build the strands of DNA that define the living organism. Interestingly, the human genome isn't just about "humans." In fact many of the strands of DNA are common across species and allows for the ability to distinguish, from a genetic standpoint, the difference between us and the bonobo.

Occasionally the genetic scribes create a typo, substituting one nucleotide for another. The body is wonderful at repairing this mistake. However, if the mutation spreads, and the host finds the alteration beneficial, it could be spread from one mutant to his or her offspring. I've been told that red heads are a recessive gene and one day, somewhere in the future, there will be no more red heads. I've also been told that my blue eyes are the result of some Russian Czar who was prolific in producing offspring sometime many moons ago.

We often associate genetic mutation with green monsters from the science fiction films from the eighties. Genetics was little understood, but the idea of genetic mutation was considered excellent fodder for the cinematic imagination. Indeed, in most cases mutation is a good thing. "Normal" corn seed (I am excluding the GMO variety and some special implementation of mono-genetic seeds) regularly mutates. The benefit is that, when a blight hits, there is enough mutant variation to provide the farmer with a barrier in the field preventing a complete disaster with the crop. The same holds true for humans. The variation prevents a strain of common cold from becoming an epidemic. The virus cannot mutate amongst the variety of genetic diversity found in a standard population to wipe out a race.

Consider a teenager in the midwest, who was in a car crash, walked away with no broken bones. Upon examination, the entire family had a genetic mutation which created denser than normal bone growth. There were additional benefits as well. The genetic mutation in this case was lipoprotein receptor-related protein 5, or LRP5[65]. The super human structure allowed the boy to survive unscathed. This genetic gift saved his life.

Italy is one of my favorite places to visit, but now that I know there is a genetic mutation in natives that reduces cholesterol, I'm more apt to get seconds on the huge place of pasta (okay - I know that is not how it works but I love Italian pasta to a fault). Italians possess the Apolipoprotein A-I mutation in protecting against atherosclerosis[66]. Understanding these traits, benefits and ability to suppress disease has changed genetic research.

In fact, a new arena of science is emerging to be able to artificially transmit these beneficial mutations. Simply changing a cell's genetic makeup is not enough. Using gene therapy, the science is looking at viruses as delivery mechanisms to replace section of DNA much like those Hox genes can rewrite the genetic story down the line.

Genetic mutation, in isolation, doesn't account for the modifications that would occur to create the coastal creatures that we've become. The random rolling of the genetic dice is insufficient to direct the course of evolution to who we are. It must work in concert with Darwin's Natural Selection. Four criteria need to be present for the continuation of a genetic modification:

- We need the mutation to begin with. Without the mutation - this entire process it moot.
- The mutation must be admissible. If the body does not accept the genetic modification, the body will consider it a foreign agent and destroy it as an invasive change, sometimes causing loss of life (and therefore curtailing the possibility of propagation).
- The lack of genetic weakness - if, during the genetic "typo" the body is able to correct the issue, it will.
- The opportunity for genetic propagation - given the prior two opportunities, the genetic mutation must be able to propagate throughout the body, and be transmitted along descendants in which it exists.

Sometime in the 1690's a sea captain and his brother settled in the, then, sparsely populated Martha's Vineyard. They brought with them a special gift to this remote island - *congenital achromatopsia*. Though they weren't deaf, by the 1900's more than a quarter of the community could not hear. When Alexander Graham Bell visited the island, he wondered whether it might not harbor an entire 'deaf variety of the entire human race[67].' Both the hearing and non-hearing community adopted a sign language. The language was particularly advantageous while sailing, when the wind howled and boat used a spyglass to transmit messages across a range that exceeded the ability to yell. Those visiting the island would regularly see the community signing to each other in town. Though this example took generations a natural disaster may funnel the genetic diversity down to a few left to repopulate, exposing and propagating the mutation in a single generation.

The story is recounted throughout history. Sometimes when foreign invaders or explorer brought disease, the introduction of these agents ended up wiping out populations. The source had built up immunity over generations and were oblivious to the crippling disease they brought. One way many island tribes survived the onslaught of "the pox", "the fever" or other European malady was to intermingle and produce children with a genetic mix of native and non-native DNA able to survive the island genocide.

Consider a small tribe in the Andes. The arid landscape bakes the skin of the people in this rural community of about 6,000 people. The Argentinian village San Antonio de los Cobres is known for its copper mines and products made by weaving the local llama wool. Some houses are without plumbing and the residents are forced to seek community water sources to help them quench their parched thirst from this desert. One caveat of this mineral rich land is the fact that their water is laced with an extremely high concentration of arsenic. They consume the water that would kill most others on the planet. Yet they are able to as a result of the gene AS3MT which allows them to process the toxin better than the average person[68]. In this case, territory alone is the harbinger of death, now averted for a select few.

A simple error in the genetic code is but one example of how the code can be changed. Indeed, a large portion of our body contains "jumping genes" known as transpositional element otherwise known as transposons. These are protein sequences that can "move" from one location to another. This is not always a destructive process. The medaka fish's transposons are linked to pigmentation[69].

These transposons play another crucial role in determining the path of evolution.

"Mobile DNA has become vital, too, in tracking evolutionary relationships among species. That's because if you compare a few species, and just two of them have the same transposons burrowed into their DNA at the same point among billions of bases, then those two species almost certainly shared an ancestor recently. More to the point, they shared that ancestor more recently than either shared an ancestor with a third species that lacks the transposons; far too many bases exist for that insertion to have happened twice independently.

What look like DNA marginalia, then, actually reveal life's hidden recorded history, and …. suddenly seemed less cute and more profound[70]."

That is why the similarity of transposons in humans is more similar to primates than birds. With this information we can begin to trace backwards through time to find when there *was* a commonality among species. Comparing fossil records with changes in DNA, we can step backwards with certainty on those traits which both unite and separate us. We can identify those strands which, for some reason, changed and were passed forward to make our toes look like they do, or drop our larynx, or make us naked primates.

The 2016 Olympics saw scores of young athletes from around the world come to compete. The spectacle was remarkable and the achievements stunning. Rio De Janeiro, normally a city of ongoing parties, blossomed in color to welcomed guests. The United States was eager to see an older Michael Phelps as he returned to the water. Usain Bolt didn't disappoint with his lightning speed on the track in the 100m. But checking the fast facts of the 2016 Olympics held another record. There were no confirmed outbreaks of the Zika virus.

Other nations were concerned because during Brazil's summer months, there was an outbreak of Zika that threatened to curtail this worldwide sporting event. Zika, a virus which can have severe effects on those afflicted and can present birth defects for pregnant mothers with the disease, is spread by mosquitos. The name Zika comes from the area Ziika in Uganda where it was discovered in chimpanzees. Five years later, it was diagnosed in humans. Similar to yellow fever or dengue, this virus belongs to the family Flaviviridae. If passed from mother to child, the child can have Microcephaly, a disease where the child's head is too small for their body. Their brains do not develop properly which can result in learning defects.

This virus, like many others act similarly to the transposons. The virus can infect the body and inject themselves into a DNA sequence. This process happens when the virus breaks through the cell wall. Some viruses have been known to develop a tube or tail-like structure to get into the cell (not unlike a sperm cell). From there it hijacks the cell and turns it into a factory forcing it to duplicate the virus, once it hits a critical point where the cell is no longer viable to the invading host, the virus exits the cell (destroying it in the process). Only if the white blood cells can produce a protein to either inform the body of the invading agent, or create a protein to bind to the virus making it unable to reproduce, will the host survive.

If the host cannot come up with a solution, the infecting agent wins and we have an extinction event. If there are a few survivors who have been able to adapt, they become the individuals who will repopulate and expand their genetic library for subsequent generations. There is a third option, however, the invading virus inject neutral RNA sequences and is not considered a threat, thus the virus is considered "acceptable" to the host to a certain point. The neutral virus may provide latent "upgrades" to the organism to be used later by a similar, but more lethal, virus. This code is passed on to subsequent generations. In fact, much of the "junk DNA" found in the human genome project is a product of invading viruses who have left their signature in the book of human evolution.

This symbiotic relationship can happen quickly. It can cause a jump in the evolutionary process, speeding up a process that Darwin could never have imagined. Patterned baldness hair loss (Alopecia Areata) may not be an evolutionary adjustment at all. One study on mice demonstrated that a retrovirus would cause similar hair loss. The retrovirus was passed from mother to child[71]. One could speculate that in a short span of time a small group of hominids encountered a virus which injected the DNA sequence producing hairlessness. This biological anomaly would be considered advantageous for coastal dwelling creatures. The lack of hair would prevent matting and diseases from collecting in the hair on the hominids' bodies. Therefore, an external force may drive evolution forward in the form of a virus that - to the host - is absorbed and replicated.

Physiological evolution which creates specialization can help a species survive. Wings, claws, venom, and a host of other tools can make the difference between life and death of an entire species.

There are other tools, however, which may not seem like game changers, but are just as important as size, weapons or camouflage. Our bipedal alteration caused a cascading series of physiological events which provided us with new tools for survival. Our vertical nature dropped our larynx and altered the location and position of our jaw. Our face changed. Our ability to articulate speech became a reality. Additionally, our brain size increased. So is there a correlation between the physical dependency to stand erect as the result of the demand to live along the coast and forage for water and an increased brain?

Without the physical alterations that took place, we would not be able to have the sized brain that we do now. Our frame allows us to balance the larger head along its vertical "pole" of a body, rather than having a set of large neck muscles to support the head like our quadruped cousins. Studies of Zika helped lead a group of researchers in Belgium to understand that there is a distinct genetic alteration that took place, unique to humans, which helped create the larger cortex.

"Three members of a gene family called NOTCH2NL may have been involved in the evolution of humans' big cortex…When the authors investigated the genomic context of NOTCH2NL, they identified four NOTCH2NL family members, three of which—NOTCH2NLA, NOTCH2NLB, and NOTCH2NLC—cluster together on the first chromosome. Their analyses revealed that NOTCH2NL-like genes exist in other primates, including gorillas and chimpanzees, and came from a duplication of NOTCH2. But the only species with functional NOTCH2NL genes are humans and their closest ancestors, Neanderthals and Denisovans. In other primates the NOTCH2NL family members are pseudogenes. The researchers determined that NOTCH2NL became functional in hominins about 3.5 million years ago after a gene conversion event that combined the promoter and first exon of NOTCH2 with NOTCH2NL. This lengthier gene was then duplicated to give rise to A, B, and C[72]."

This means that the gene exists in other primates in some altered or dormant form, but in humans became active as the result of a need for a larger brain. But in order to sustain the alteration, the environment would have to be able to promote our ability to sustain this change, otherwise the alteration would die out. In other words, whether as a result of our change or a change placed upon us, the alteration did not die out because it was considered "deficient" either by internal or external forces.

It is extremely important because in order to survive with a larger brain there are three nutritional requirements - a long-chain, polyunsaturated fatty acid – *docosahexaenoic acid (DHA)*, *arachidonic acid* (AA), and *eicosapentaenoic acid* (EPA). DHA is an omega-3 fatty acid is required for development of the brain and "has a positive effect on diseases such as hypertension, arthritis, atherosclerosis, depression, adult-onset diabetes mellitus, myocardial infarction, thrombosis, and some cancers[73]." EPA is the sister protein to DHA, required in much smaller quantities, and provides similar advantages as DHA and both are often found together in fish oil health products. However, they serve different functions and act on the body in different ways. EPA specifically helps with cellular inflammation. The more EPA you have, the less AA you produce. EPA and AA are like a two cops looking after the human brain and nervous system. When one is on the job the other can rest and vice versa.

So anthropological evidence which demonstrates an increase in brain size must have a correlated increase in the requirements of these omega-3 and omega-6 proteins. Those still adhering to the Savannah Theory have no footing for the dietary needs of large brained hominids. The grasslands do not provide the sustenance for the hominid brain. The coastal wetlands, by contrast, are rich in fatty acid proteins found in both fish and shellfish. A coastal environment, therefore, is the answer for how humans could survive, thrive and evolve with a larger brain.

Scientific studies comparing body/brain size conclude that we, once again, are the anomaly. The results show that as the body size increases, the brain size *decreases* logarithmically.

A Body of Water

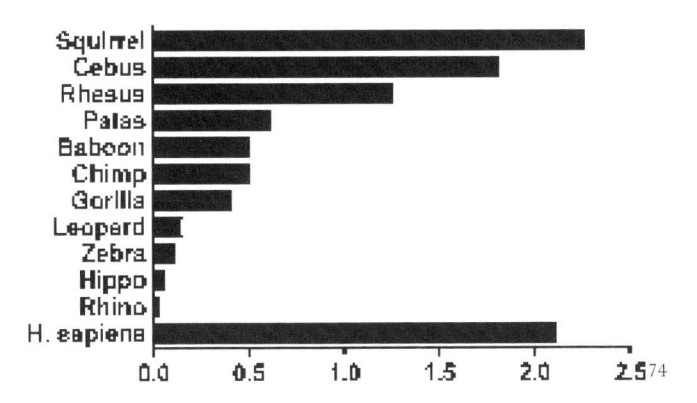

Animal	Value
Squirrel	~2.3
Cebus	~1.8
Rhesus	~1.4
Palas	~0.6
Baboon	~0.5
Chimp	~0.5
Gorilla	~0.4
Leopard	~0.1
Zebra	~0.1
Hippo	~0.05
Rhino	~0.05
H. sapiens	~2.1

2.574[74]

So we are left with a conundrum unless we factor aquatic animals into the equation. "Of the large mammals, the dolphin with about 1% of its body weight as brain comes the closest to *H. sapiens*[75]." This is not to imply that we were, at one point, akin to dolphins. We've never been mermaids, cetaceans, mollusks or anything other than the primate that we are. So often, the comparison of body mass/brain weight to dolphins causes the scientific community to roll their collective eyes. The important point to note is that, given this similarity and the nutritional requirements for proper brain development, hominids required a diet that included those things found readily along the coast.

However, the human evolution as tied to evolution may occur in a single generation and not as a long slow progression. Whether we look at people able to consume arsenic like the tribe in the Andes, or the Philippine boat people who can hold their breath, beneficial genetic drift is now understood to have played a crucial role as we left the tree tops for a dip in the cool waters.

At some point, however, we crossed some invisible threshold when we left our treetop homes in favor of the land. A small change here, perhaps the change in the anatomy of the foot, which led to the ability to sustain an upright stance. Then another change here, and another there. Until, at some point the framework that creates a human being was altered to the point that we would not - could not return to our arboreal canopy. We became something so totally different. Humanity found its way to the shore and never looked back.

BEACH HOUSE

The first rule in hurricane coverage is that every broadcast must begin with palm trees blowing in the wind
— Carl Hiaasen

An ongoing problem with the Florida coral coast is the introduction of the lionfish (Pterois). This creature who is found in the Pacific asian waters has made its troublesome home in the Atlantic. It is prolific in reproduction and the underwater community of animals has yet to identify this fish, with its poisonous needles, as a source of food. It is prolific with females able to generate 2 million eggs a year. Other than humans, lionfish reproduces mostly unhindered. Humans are the primary caretakers, trying to limit the spread of this invasive species.

The problem is confounded by large tankers traveling the world, transporting its wares everywhere on the planet, which at a port of call may take on water as ballast in Asian waters and upon arrival in the United States release the water and anything contained in the marine storehouse when they arrive at port in Florida. Until nature catches up, these invaders will continue to proliferate in Caribbean waters at an alarming rate.

In this case, we are the harbingers of environmental change which introduced this planetary dilemma. Until nature somehow identifies the solution, it will continue to repeat unabated. This is but one alteration in the world which has occurred as the result of man the likes of which I've witnessed growing up. Other factors have altered the fabric of the planet. Floods, earthquakes, global warming, and drought alter the drama of life to this day.

As the result of climactic changes about five million years ago, early hominids moved from the forests to the coast. It is likely that the change in the sudden appearance of these small hairy beasts weren't noted by the community of both aquatic and terrestrial predators, leaving these solitary hominids to thrive unchallenged. While the early humans adapted to their new environment, they grew in number and gained skills to improve their survival. This grace period likely led to the vertical ape that is our ancestor.

Humans who lived along the water's edge had the benefit of the protection from both environments. Those that might prey on humans would have to be adept in both environments. Though it is well known that alligators and tigers can visit their neighbor's environment, they lose much of their advantage in the process. Alligators on land are exposed and not as quick as in the water. Tigers, while swimming, lack the capacity for a speedy attack. A human that can quickly transition from terrestrial to aquatic may gain enough advantage to survive. Edge dwellers can take advantage of both environments. During the honeymoon period five million years ago, the new inhabitants could not only populate, but begin to hone their skills in their newly adopted environment.

And, at some point, with learned proficiency for survival and a developed routine for daily life, the environment became *territory*. This is something that you claim. This is something that you own. The territorial nature of man takes over and you defend and tend this area. It is here that community is formed. This goes beyond mere dwelling and becomes home.

When we go on vacation to a beach house or some other place which we temporarily inhabit, we take only this things we need for the stay. We use whatever resources are at our command, knowing that we will be returning to our home. We don't bring a washing machine for our clothes. We don't bring an ironing board in our suitcase. This is a temporary dwelling. However, when we buy a house, these things and countless others become the necessity of living as we define the territory that has become our home.

Patriots defend their country, farmers defend their land, and families defend their home. It is often overlooked that the desire for territory extends beyond even our desire for sex/reproduction. We fight and scrap for our place within society, wherever that niche places us.

If a territory is rich in resources, that space becomes the prime real estate for individual in the upper echelons of society. Given the complex and structured hierarchical nature of every primate ever studied, it holds that territory would be reserved for the alpha-group in the community. If a particular individual or group has sole access to a given set of resources, we suddenly move from *territory* to *property*.

Property has a slightly different connotation than territory. Territory is owned by the community while property implies ownership by the individual. The individual (or individuals who share ownership) must secure the rights to that piece of territory. In that sense they retain possession of the property, claiming it, defending it and nurturing it as needed.

But if we own *property* we do so within the confines of the larger *territory*. That includes not only the physical resources, but the tribal requirements of the population in which we inject ourselves.

"What seems to set human property apart from that of other species is the extensive reliance on the goodwill of others to assist in the maintenance of ownership (e.g. property outside of one's possession) through third-party reinforcement. In humans, this takes the form of both institutional structures to maintain ownership rights (police forces, legal systems) and the tendency of humans to respect each other's property ownership. As mentioned above with respect to the ravens, third party norms do exist in other species, but typically only for current possessions, and not with respect to ownership. This may be due to the inability of other species to convey information beyond the immediate, as can be done with language. This means that third-party interventions can only occur in situations in which the transgression was witnessed by a potential supporter (Brosnan, Grady, Lambeth, Schapiro, & Beran, 2008). Moreover, while many of us may resort to the legal system to reclaim property which has been taken from us, the truth is that in a well-functioning society, this recourse is required surprisingly rarely, particularly with respect to how often property we own is left outside of our immediate possession[76]."

Primates have a very loose sense of ownership, basically covering food in possession of the individual at that time. We are unique in our sense of ownership among primates, for the most part. Territorial disputes are commonplace. On any given morning in lands with the howler monkey, you will be awakened by the air-raid like hoots and hollers of these tree dwelling primates. They howl with volume to humble any opera singer. Each howl indicates the area and boundary in which a particular tribe is now inhabiting. Should a howler monkey howl along an edge of another's territory it is taken as a threat and the collective primate-police head to the area of infringement.

A friend of mine who lived in the dutch region of Pennsylvania woke up and yodeled. This was a daily occurrence and each family had a different yodel unique to their "tribe." Should an individual fail to yodel, the rest of the neighborhood (spanning many acres), would come to see what was wrong.

Within the larger context of population of any species is the desire to carve out the niche for their "own personal space." One merely needs to look up at a telephone pole filled with birds to see the space that separates them, each arranged like chess pieces, equally spaced, and distributed with mathematical precision. The space is not so great that one cannot manage their territory, but large enough to satisfy that inward need each of us feels for our flight or fight zone.

On the fringes where human population is sparse, we have to be self-reliant. One problem that anthropologists face is the misunderstanding of cultures as *primitive* equating to those who are sparse in number. Whether it is the Yanomama Indians or the Inuit of the Alaskan tundra, anthropologists run to these examples of *primitive* cultures, which - in fact - are not primitive at all, but rather represent a specific quality of life. The point is that primitive and self-sufficient are two different cultures, not to be confused with one another. Large packs of animals represent one state and the "lone-wolf" represents another. Primate social structures, as mentioned previously, are quite complex, originating in its earliest form as parental division. Hominid Coastal living would require foraging in two environments and establishing protection from different threats requiring specialization.

Associated with property, ownership and securing the rights to land, is the division of labor. In a mono-environmental landscape (say a tree canopy for instance) it is easy to share responsibilities because the methods of securing it is the same from edge-to-edge. However, in a coastal environment we have both land and water which needs to be secured. Because they may possess differing threats, specialization becomes necessary. The division of labor takes on greater and more complex variety against scenarios coming from water or land. This requires greater coordination among the group. Society becomes more complex as each group within the larger tribe performs their assigned roles.

Whether we like it or not, our striving for individuality is built within the larger framework of this division of labor. We all do not create and consume our own food. We do not generate all of the means for our life support. We do not care for the refuse that we generate. Indeed, as the size and complexity of our culture expands, we play a smaller slice in the larger picture of the society that is human. Yet, at its core is the requirement that we participate in the same role that our ancestors played as one worked the land and another hunted along the coast. We, in our desire to find self-actualization, are localized within our tribe to conduct our part no different than that set forth millions of years ago.

As my friend Will once said, "We have a right to be different, just like everyone else." I wrote, in one of my novels that a character hoped to win the lottery so he could live the simple life. Our means may not justify the end, but the goal is the same: to follow that instinct which brings our body and mind into harmony. But we do not make that decision alone. The devil and angel on our shoulders poking and prodding us one direction or the other are related to us. They contain the whisper of our past. To infer that we have lost our base instincts is to deny who we are at our very core. If we adopt Maslow's hierarchy, we cannot deny our instincts as much as if we were to build a two-story building and claim the second story bedroom and deny the foundation and first floor. In the play of our life of modernity, we often ignore those wheels within wheels that drive our decisions, influence our perception, and whisper in our ear. Lest we forget, we are animals. I fear to add the term intelligent, for one needs to ask what that really means. If it means living in harmony with our environment and each other, then we are sore losers in that game. We are (mostly) sentient creatures, but we are creatures nonetheless.

A bird raised in captivity will still sing the songs of its kind. Living within each of us the ancient echo of our distant past. Whether written in the genetic code or passed down from parent to child, the conditions that formed us span generations. To that end we still harken to the calling of our past. But what is that calling? Where are we drawn? Can we still hear the echoes that reverberate over such a long span of time?

If we look to ancient texts, that refer to even more ancient times, water is crucial in the myth stories that emerge from multiple cultures. I am using the term *myth* in perhaps a way that the reader may not be accustom. To understand myth, we need to see this as a tradition that is passed down either orally or written, to convey an ideology of a people, through law, instruction and allegory. Whether we refer to Gilgamesh and Noah, Genesis or the earliest writing of Mongolian tribes, water is critical in creation (and destruction) within the cosmos.

Specifically devastation stories often surround an ancient history or allegory of an event or events related to the actions and activities of a local population in relation to their environment. One has a hard time distinguishing the two because the land (or part of the land) and the water is often portrayed as an individual or individuals as Gods or, more often demigods. It is through environmental change that these myths wreak havoc on the human population. But if one reads the stories backwards, one reads the environmental consequences wrought by deviation from the tribal custom.

We need to rereads these stories from an eco-theological lens in order to better understand the implications for them - and for us. Embracing and understanding the myth is crucial for coastal societies and beyond. Edward Evans Pritchard, an anthropologist stated:

"... Nonbelievers would never come close to understanding religion and myth as well as beliefs. Nonbelievers tended to try to explain religion away as illusion, using sociological, psychological, existential, or biological theories[77]..."

Consider Genesis which has two creation stories - one which is water-based and the other which is land-based. Gen. 1 holds an orderly creation over the deep waters and Gen 2:4 is a narrative of land based creation. This holds that in the ancient stories of creation, both stories are held in equal importance and value such that they both are represented in the Judeo-Christian doctrine. Scientists would downplay the fact that there exist conflicting stories within a single tome, rather than absorbing that each is valid and important in the cosmological story.

Moana, a Disney production which took some liberty with old Polynesian myths, portray an intimate relationship with land, water and their interdependence. Moana—pronounced "moh-AH-nah," not "MWAH-nah" means "ocean"—and the character is chosen by the sea itself to return the stolen heart of Te Fiti, who turns out to be an island deity (Tahiti, in its various linguistic forms, including Tafiti, is a pan-Polynesian word for any faraway place).

The heart of Te Fiti is a greenstone (New Zealand Maori) amulet stolen by the demigod Maui. An environmental catastrophe spreading across the island makes the mission urgent. And despite admonitions from her father against anyone going beyond the protective reef, Moana steals a canoe and embarks on her quest[78].

"Kapu" or taboo, was a law instituted by a chief which prohibited activities. If the law was broken, this could result in death of the offender. Often Kapu was instituted against certain fishing practices for a period of time to maintain proper balance of stock. In a sense, the chief of the Polynesian island would preside over the health of both the people and the land.

"Over a period of many centuries the Polynesians who inhabited Hawai'i developed a carefully regulated and sustainable "ahupua'a" management system that integrated watershed, freshwater and nearshore marine resources based on the fundamental linkages between all ecosystems from the mountain tops to the sea. This traditional scheme employed adaptive management practices keyed to subtle changes in natural resources. Sophisticated social controls on resource utilization were an important component of the system. Over the past two centuries a 'Western system' gradually replaced much of the traditional Hawaiian system....The predominant traditional system in the eight high islands of the Main Hawaiian Islands (MHIs) was based on the ahupua'a, which is a unit of land that extends from the mountains to the sea and generally includes one or more complete watershed(s) and all nearshore marine resources [19, 20]. Each ahupua'a contained a broad cross section of island resources and was managed within a complex social system associated with each area. The general belief is that each ahupua'a met the needs of the local population with an excess for tribute and trade[79]."

These western practices (read world-wide consumption) policies have left coastal cultures decimated with little recourse other than to either

- Create a dependent alliance on the offending, intrusive agent stripping the coast of its resource
- Challenge the intrusive agent (often with limited resources) usually leading to the decimation of the coastal society
- Migrate to another location and claim the fertile territory (which may be inhabited by another population)

- Identify alternative resources for survival (untapped or unclaimed resources) - usually hastening the coastal environmental decline

One coastal population driven to piracy is Somalia. The chaos of Somalia is but a preface to larger coastal towns across the world suffering similar fates. Somalia, once hailed largely as a fishing population, political infighting blinded them to the villains of international fishing that was raping their ecological resources. Specifically Russian fishing boats would skirt the international waters, often venturing inside the abstract line of demarcation to gather what they could with long-lines and trolling nets - both equally devastating to the ecology. Not realizing it until it was too late, these local fishermen now turn to piracy as their means of survival. With the coastal ecology crushed, they have little other resources to offer.

Each year Japanese fishing has plummeted, much like the cod fishermen of the Northern Americas. What is left is an odd array of jellyfish which pollute their nets and make dangerous the separating of their viable catch from the gelatinous detritus.

I had the pleasure of a few hours with two gentlemen from New Zealand, stopping over for a few hours on their way home from the Marshall Islands. These lost islands claimed by the United States are following the same fate as Somalia. Tribal infighting is covering the larger problem of dwindling resources.

Those who adhere to the "old ways" of fishing are up against the illegal, but little enforced technique of dynamite fishing. It is basically what you would think it would be. A boat goes out with a hoard of illegally obtained dynamite and stuns the fish to the surface (while destroying whatever aquatic landscape may be flourishing on the bottom). The catch is harvested with large nets. In many cases the boats never stop in a single location and merely drop their explosive, perform a single rotation around the point of destruction and move on.

Worse, those tribal fishermen have been known to come home and seen their processing plants blown up by the very same dynamite. The police are helpless in identifying the culprits who are often transient fishermen from other areas who disappear the moment their boat becomes suspect.

The gentlemen were sewage experts who, upon arriving to Marshall Islands, determined that the destruction wrought by the dynamiting of the facilities also destroyed a portion of the underground sewage system, thus allowing raw sewage to run off to the ports and coastal areas where the greatest number of inhabitants (many now poor from overfishing) lived. In fact, many of the indigenous people were being relocated inland by the government to help stem the tide of poverty. These newly constructed housing had no means for support for a people accustomed to coastal living, now forced to live and create a life with no retraining for providing means for their families.

As I spoke with these gentlemen, over a Mai-Tai on one of the Hawaiian islands, I couldn't help but recall the song, *Here Comes the Flood* by Peter Gabriel:

Lord, here comes the flood
We'll say goodbye to flesh and blood
If again the seas are silent
In any still alive
It'll be those who gave their island to survive
Drink up, dreamers, you're running dry.

We are drawn to the coast. Approximately 80% of the world's population lives within a drive to the shore. One reason is that the coast highlights a boundary between two very different ecosystems. It also demonstrates the interaction and dependency these ecosystems have on each other. Whether it is the elimination of single-use plastics or the policing of sunscreens, coastal communities seem more aware of the need to "nurture nature."

"When Gordon Jones, owner of Seaside Realty and Seaside Vacation Homes, on the Outer Banks of North Carolina, surveyed thirty-five real estate agents to discover why their clients had bought oceanfront homes, the answers included:

- To hear and fall asleep to the sound of the ocean
- A status symbol
- The pull of the ocean, the serenity and respect

- The ultimate challenge of mother nature
- A good investment/rental income
- Lifestyle
- To watch the sunrises
- The convenience of having the beach at your front door
- Inspiration - people are moved to write, paint, or do whatever they do best
- To see wildlife - pelicans, whales, dolphins, sea turtles, fish[80]"

Small communities whose lives depend on tourism tend to maintain strict control over waste, plastics and activities detrimental to the ecosystem. One evening while taking a solitary sunset walk along a beach in St. Augustine, I came across a group of people gathered as a turtle nesting site hatched. The moment was coordinated by a few locals who kept the tourists from damaging the nests or interfering with the turtles as they made their way to the ocean. We all became cheerleaders at a pep rally, filled with childlike glee as these little ones emerged from their sandy lodging to meander toward the water. Places like St. Augustine have adopted lighting practices to prevent light sources at night from luring these turtle infants away from their watery destination.

In a sense, these resemble what might be considered an abstract form of coastal farming. This may stray far afield from the traditional notion of farming, but if these small coastal populations are tending to the environment in an effort to grow "tourism" then they represent conscious choices toward improving their land/water symbiosis.

Of course, modern opinion for living along the coast could be viewed as merely choice for a lifestyle and social status. The science of choice is rather nebulous. But if price is an indicator of desire, then the coast is the obvious winner. If you look at the speed with which real estate sells, the coast is again the winner. Poll after poll identifies the coast as the desired location for the majority of the population.

But that is just part of the picture. One must ask, why do we *want* to live along the coast? Is there a psychological reason? Is there a physical reason? Can we apply a scientific method to understanding the *desire* to live along the coast?

Immersion in water produces an interesting change in the human body. As mentioned before, the mammalian dive reflex causes physiological alterations in the heart, kidneys, etc. One study placed a test group immersed in water for hours. The result was a 61% drop in norepinephrine.

The presence of norepinephrine increases alertness, anxiety, and fear. It also promotes an increase in blood pressure. The lack of this substance can induce meditative states and a sense of calm and peace[81]. It appears that being in water is good for us on many levels.

"Aquatic activity impacts the cardiovascular, musculoskeletal, autonomic nervous system (ANS) and endocrine systems in ways that have positive public health implications for issues confronting the nation, including obesity, diabetes and arthritis (Becker, 2004). Aquatic activity has tremendous application in the area of sports medicine and has great potential value to student athletes in both training and rehabilitation. The aquatic environment is a research area just emerging as a focus of physiologic importance with many health benefits that apply across the age span and could be widely accessed by the American public if both research support and under- standing by the health professionals were to increase[82]."

Woody Alpern, a SCUBA instructor regularly talks to his students about the recuperative processes found when one plunges beneath the waves.

"There is a life-force - an energy that exists in the ocean. It fills you up and fixes you in ways no other can. After I spend time in the ocean my batteries are recharged. It is indescribable, really. You have to jump in and experience it for yourself. Only then can you really know the magical power that the ocean has for each of us."

Woody has spent a good deal of time in the ocean. As both an open water (regular SCUBA tanks that most of us are familiar) as well as "closed-circuit systems (no bubbles - the system recycles the air you breath making it more efficient for diving to depths and times greater than normal recreational diving) Woody has travelled the world and considers himself a coastal apostle.

"People know how to dive. I am not so much teaching them as allowing them to reveal what they already know. A large part of my time training a student is to simply let them play underwater. If you let them figure it out early, they'll be much better divers. I've seen other instructors who overweight their students. They send them to the bottom of the pool and get on their knees to conduct the training. That's not normal. We are water creatures. If I can't teach them how to perform the tasks while in a neutrally buoyant state, then I'm not doing my job. But, if we let the body take over and stay away from the mechanics that get us in trouble - like over inflating, overweighting, etc - we produce better SCUBA divers and engage the student in a more natural process of diving."

Speak to any apnea diver, SCUBA diver, surfer or paddle boarder and you begin to hear the same poetic, zen-like qualities and words rattle off their tongue. There is a state of oneness that one obtains in commune with the blue abyss. There are revelations that transcend scientific inquiry. There is a secret society that one has to experience to enter into. For each specialty, each monastery or sanctuary of the aquatic divine, there is a lingo and tribe which is both difficult to understand and participate. Whether we are talking about surfer "breaks", apnea CO_2 tables, or SCUBA DCS, each coastal group maintains a tight set of hand signals, verbiage, and culture unique to that location and group.

But a commonality among them is the tie that they have to both land and water. They have a greater commitment because of their interaction and witness to the state of the health of the environment in which they live, thrive, and find enjoyment. Sylvia Earle, a long-time SCUBA diver, proponent for ocean health and scientist said, "Why is it that scuba divers and surfers are some of the strongest advocates of ocean conservation? Because they've spent time in and around the ocean, and they've personally seen the beauty, the fragility, and even the degradation of our planet's blue heart."

Those visitors to the beach community, SCUBA resort, or coastal destination comes to feel the change in self. For some it is simply to be able to look at the stars. It is a way to reset the circadian rhythms that we lose in an urban, technologically charged environment. Science has proven that even a visit to the coast will change us. Whether it is a lowering of the toxicity that binds us when we are away from water, or an alteration in our psychological outlook, for most, the draw is palpable.

The question is whether we can draw a definitive line from where we are now, to that moment when we left our primate cousins to take to the water. The physical evidence for a coastal existence still permeates our lives today, from our stance, to the way our body reacts to immersion. Many in the scientific community still refuse to see what, for many of us, is an obvious conclusion: that we are coastal creatures whose very difference is the signature for what happened.

WHEN OPPORTUNITY KNOCKS

Alternatively, anyone who favors Intelligent Design in lieu of evolution might pause to wonder why God devoted so much of His intelligence to designing malarial parasites.
— David Quammen, Spillover: Animal Infections and the Next Human Pandemic

The conundrum that plagues scientists arguing against the Aquatic Ape Theory (Hypothesis) and the Coastal Theory is the basic premise of environment. The scientists still clinging to other theories use the savannah as the theater upon which the theories fall. If the set is wrong, then humans never lived along the coast and subsequently, we never became bipedal as a result of our new found home-sweet-home. Like the unraveling of a mystery novel, our ancestral homeland plays a key role in determining the 'whodunit' of the plethora of hypotheses that exist trying to tell the definitive story of our lineage.

It was a challenging period for me as I lay, for weeks, in the hospital. I was diagnosed with cancer and was recovering. I was desperate to get out. I was off the critical list. So they were going to move me. Luckily, for the hospital, a guy down the hall just died. That was the room I was going to occupy the next day. I dreaded the thought. During my short stay I met a girl named Amy. Her boyfriend was with her. The large milky substances hanging from the metal pole, providing her with a continuous flow of chemicals, never left her side as her boyfriend wheeled it when they took their afternoon walk. I told her I was leaving in a week. She told me she was leaving in a month. I congratulated her. However, when I looked at the boyfriend I understood the gravity of her statement.

All was going according to plan until I moved into my new dreaded room. It still smelled of death. Perhaps that was just me, but I felt an ominous foreboding. My intuition was confirmed when, before the end of the day, I started getting sick(er). The doctor determined I inherited *Clostridium difficile* a.k.a. C. Diff. It tore me to pieces and nearly took my life.

This is but one of the many super bugs that hospitals contend with on a daily basis. Everyday new bugs emerge to rattle the medical community. Like weeds cropping up in a human garden, we are at war with an ever changing medley of destructive bugs threatening the human race.

Penicillin was touted as the WWII miracle drug. That was, until Staphylococcus became a victor, Tetracycline was the next drug and lost out in about nine years. Methicillin had only two years for a victory lap. Observing evolution on a microscopic level, we see a rapid and unnerving escapade of virulent foes arriving, or reemerging (like the measles) with increasing rapidity and strength. So how long did it take for us to become "human?"

We are also dealing with timespans that most have no idea how to comprehend. From the microscopic day to a million years, our ruler upon which we base the context of our theories wreaks havoc upon our analytical discernment. Speculation on which we play this game may be off by tens of thousands of years which, both in terms of ecological consequences and the human genetic factor is an immense margin of error. It is like being invited to a party but you aren't exactly sure if the party is Saturday or Sunday. The time factor is incredibly important. It is also in this incalculable fodder that scientists stake their ground and pound the table in disbelief of one theory or another. Sadly, we are all stating conjecture on our best guess, and that's it! Perhaps by me announcing forthright that I am doing my best, but this is a painted scenario rather than some demonstrable truth I've given up my authority. However, I'd rather invoke the truth at this point and let you, the reader know that I'm doing my best, but I'm doing no better or worse than others publishing in scientific journals on the same subject.

We can look into the crystal ball and pick a random point in time, say five million years ago, give or take a few thousands of years, and look at the evidence from that time period. We can analyze the dirt to determine if it was a wet or dry period, we can scrape around for bones, both human and animal, for evidence of life. If we are lucky we might be able to render a DNA fragment. All of this helps build the world in which our ancestors lived.

To grasp the importance environment plays in our evolutionary upbringing we need to look at just about any other animal on the planet. The animal kingdom is replete with breast beating and territorial protection. Suffice to say that animals identify their homeland and will defend it fiercely. Moreover, those boundaries that are abstractly laid out like the divisions of nations in the human world, also exist in the animal kingdom.

Look at the map of the United States and read it from right to left. No, not the names, but the borders. It demonstrates a social evolution that occurred as the United States marched westward. The older borders followed natural boundaries. Maryland's southern border is the Potomac River. Whether the Delaware River, the Chattahoochee, the Ohio River, or the Smokey Range these landmarks helped define the early formation of states. Moving west we see a change as society advanced with the industrial revolution and the rise of scientific thought. Moreover, we see a species dictating territory, not by the physical limitations preventing us from leaving one space and entering another, but rather we see a division based on politics, society and the democratic action of border drawing. Not unlike animals who thrive in the wild, they too will stake their claim, whether it is howling their dominance from the treetops or urinating on the territory's edges.

Allopatric speciation is the common lens through which we gaze into our evolutionary past. The natural borders of Africa lent to the division between us and our predecessors. Flood or drought? Forest canopies or mountain ranges? These external boundaries have been the means for our analyzing the cause and result of everything we are from bipedalism to dietary choices. Nature, time and time again has dictated allopatric speciation as the root for evolution. It fits in nicely with the Darwinian theory.

"The Sanaga River forms a natural boundary between Nigeria–Cameroon and central chimpanzee populations whereas the Congo River separates the bonobo population from the central and eastern chimpanzees." I knew of the latter division. The former was novel to me. In fact I'd never even heard of the Sanaga river prior to this research. Though the Congo seems clearly a significant geological and hydrological entity, I'm not quite so sure of the Sanaga. The division between the chimpanzees of Nigeria-Cameroon and those of the western Congo region may be one with an overdetermined number of causes. Nevertheless, taking these riverine features as given parameters in generating allopatric speciation and subspecies level differences[83]"

Once again, I'm reminded of the neat displays in the museums. Scientists like to put things into orderly boxes, but evolution is a messy process, filled with quarks that derail the equations and theories. This is one such case. We revel in nature's redesign when isolation occurs. But focusing on Darwin we see one fallacy in the scientific process when it comes to isolation. It is an easy mistake: Geographic isolationism equates to reproductive isolationism. If circumstances dictate differently... if we are in a dry spell, suddenly we have cross pollination of species who diverged and now unite. But that is not how this works. Animals are highly territorial and, even when the boundaries are abstract, drawn by alphas in the tribe, sexual reproduction rarely crosses those invisible boundaries. Even more rare - once divergent groups reunite, it is more often with hostility and not intermingling. And, there comes a point where the separation is so grand (over the span of thousands, and perhaps a million years) that true speciation occurs and the two groups do not even identify with each other as remote cousins.

What if it wasn't a flood? What if it wasn't the forest turning to desert? What if "man" simply decided to come down from the trees?

Speciation can take interesting turns which speeds up the process from thousands of years to a single generation. Take, for example, Britain's and Ireland's peppered moth. This common moth becomes invisible when it lands on lichen covered trees. It's random spots meld into the multicolored background. Its history took an interesting turn during the industrial revolution. Within a few years the peppered moth disappeared, or so scientists thought. With industries belching out soot and pollution, the trees upon which the peppered moth found became black. And so, too, did the peppered moth. It had to adapt quickly because its camouflage no longer worked. And, when these countries cleaned up their act, the peppered moth returned to is former multicolored glory[84].

So, did the peppered moth identify its problem and ajust accordingly? Or, did the predators take care of culling the squad of improperly colored moths? Today, bees in urban environments are using plastic in the construction of their hives. Swans in England are making nests out of garbage. Those who can adapt survive, those who cannot perish. Most scientific inquiry describes the various processes that occur to build the case for how species change. But little time has been given to the *why*? Why did we descend from our arboreal homes and descend to find our abode on the ground - ill equipped and ill trained for this new space? By not asking the *why* scientists often fail to postulate the source of their theories. And the *why* has to transcend external circumstances feeding the end result. It may have been internal.

One thing that is little changed from our ancestral path is the order and construction of our teeth. We have an omnivorous mouth. What if the result of our evolution was a modification in dietary preference? Most larger primates are omnivorous. Lucy, the famed Australopithecus was found near crab claws and alligator eggs. It is well known that chimps, bonobos and orangutans are opportunistic eaters and will eat eggs when they come across them. If we chose to come down out of the trees because of culinary choice, then all other practices for hunting and foraging would follow. In this case we aren't looking at the evolutionary path by *allopatric speciation*, we are looking at the birth of humanity via *sympatric speciation*. Two groups would share the same space, but split based on lifestyle. The "other" would be those who remained in the treetops while we foraged along the waterlines.

This, I will admit, is rather far-fetched. The question, though, is simply the division of land and arboreal enough of a distinction to generate a new species.

"…geographical isolation must precede all other forms (reproductive, ecological, behavioral) of isolation. Maybe the geographical boundary is virtually undetectable, but it has to be there. It has to block the mixing of genes between two groups—at least initially, until behavioral or ecological differences arise that are themselves sufficient to block gene mixing. One implication of this view was that field scientists should carefully reexamine those putative cases of sympatric speciation.[85]"

And we need to consider this split, however it happened as two separate events. Speciation and evolution are two different things. Phyletic evolution is the dividing line between two species as a matter of time. It sets apart the chimpanzee and the bonobo, though visually similar as separate and distinct. Speciation looks spatially across distance. So what makes one a separate species? At what point do you cross the threshold from mere variety to a true separate entity, distinct from the "other?"

Alfred Wallace, a contemporary of Darwin discovered the theory of evolution as well. In fact, had it not been for a series of mishaps, he would have published his findings before Darwin. The twenty nine year old scientist had been making his money collecting birds of paradise to fund his solitary journeys into the wilderness. After being considered lost for two years somewhere along the Amazon, he reappeared.

Wallace carted his insects and birds through the jungle. He kept meticulous records, despite the primitive - almost impossible - living conditions. He suffered yellow fever among other maladies. Suffering bouts of depression, he never lost his scientific rigor.

In fact, Wallace's approach to taxonomic tagging was a step up from Darwin's. During his disastrous and life-threatening tour in South America, Wallace tagged the birds and beetles he collected, not only with scientific naming, but also geographic location - a point that eluded Darwin. It would prove advantageous in that he was able to identify location as a primary factor in speciation.

He'd spent his life savings to conduct this scientific journey, hoping that the collection he'd amassed would bring him profit by selling it to a Victorian society curious to purchase oddities from far away. Of course, following his cursed life, Alfred Wallace woke up by the cries of the boat captain along the Amazon, to discover that the hold containing all of his specimens was on fire.

As Darwin was preparing his presentation for the Linnean Society in Burlington House, Piccadilly (which is now part of the Royal Academy), Darwin received a letter from Wallace:

"This summer will make the 20th year (!) since I opened my first-note-book, on the question how & in what way do species & varieties differ from each other."

Darwin was "smashed" (his word). It was that letter which kicked off the proverbial "paleontological space-race" between the two men. Darwin was aware that he need to make his mark first and, true to Wallace's curse, Wallace was laid up in Malaysia. The two continued to correspond in a cordial fashion, and Wallace acceded the claim to Darwin as first, but it was Wallace who actually pinpointed both phyletic evolution as well as allopatric speciation as marks for generating distinct animals.

One conundrum Wallace faced was whether his collection housed sets of distinct species, or a set of common animals who simply possessed alterations in composition as a result of isolation. His notes on the subject help identify the difference not from an observational standpoint, but from the viewpoint of the animals themselves.

"The constant preference of animals for their like, even in the case of slightly different varieties of the same species, is evidently a fact of great importance in considering the origin of species by natural selection, since it shows us that, so soon as a slight differentiation of form or colour has been effected, isolation will at once arise by the selective association of the animals themselves[86]."

Following Wallace's definition, when did "we" become "us" separate from "them?" What were the factors that determined those differences?

Some would argue that it is our intelligence and ability to reason that differentiated us from the "lower" animals. This is the worst kind of anthropomorphism where we both construct the criteria and become the judge upon what these two terms actually mean. If we define intelligence as one who lives in harmony with their environment then we are running close to last in that assessment. If we look at socialization and communication as a means for determining intelligence then cetaceans have us beat hands-down. If we look at construction, I think the ant world and bees have us there. Lest we skew our results in our favor, this criteria is a poor expression of our differentiation.

Another might argue that it is self-awareness. This again is an anthropomorphic abstraction bent to make us worthy when, in fact, the criteria is foggy at best. In fact, New Zealand passed *The Animal Welfare Amendment Bill* in 2015 which declares animals as sentient beings. Under this bill they can fine companies $500,000 (U.S.) for animal testing and cruelty. In 2013, India's Ministry of Environment and Forests agreed to ban the use of dolphins and other cetaceans such as whales and porpoises for public entertainment under the guideline that they are considered sentient non-humans. In May of 2018 Canada will be voting on a similar bill named Bill S-203, or the "Free Willy" bill claiming these are sentient beings. In the United States lawmakers have proposed similar bills, claiming that the sentient beings are being held as hostile captives and should be granted the same rights as prisoners of war under the Geneva Convention. To claim that we are the only sentient beings on earth is not only wrong, it is foolhardy.

Baring these explanations as what truly defines humans, we are left with the physical properties that make us unique. The single differentiator that we have returned - again and again - in this text is primate bipedalism. Bipedalism set the clock in motion for all other modifications that led us to the imperfect vertical creature that we are today. Bipedalism altered the construction of our foot, it altered the placement of the head on our body. It changed the way we talk. It changed our back, our hip and our neck.

If bipedalism is the cause, then why did we ever rise up and never resume our quadruped stance?

Tying speciation to bipedalism is only tying the symptom to the disease. The root cause has to be something else. Returning to the throwing theory, the heat dispersion theory and a gamut of others already discussed does nothing to determine speciation. In all of the other theories, there is no discussion about how we became "human." By decoupling bipedalism and speciation we never arrive at a satisfactory answer for the true *why* we became human. All of the other theories rely on sympatric speciation as the possible outcome. However, as most scientific texts will demonstrate, that is a rare moment and, usually upon further inspection is found that some delineation occurred to create the two separate species.

The only theory that makes sense is the Coastal Theory.

Isolation is usually the result of impassable geographic boundaries. In the rare case it is tribal separation, but in this case we still find the odd interloper entering foreign territory as a lone opportunist. In the case of drought, forests can become patchwork quilts. In this scenario animals will tend to congregate in tighter squads for the diminishing resources. If the population exceeds the resources, the result is not usually relocation - it is extinction. Pick your point around the world.

"Now take the same logic outdoors and it begins to explain why the tiger, Panthera tigris, has disappeared from the island of Bali. It casts light on the fact that the red fox, Vulpes vulpes, is missing from Bryce Canyon National Park. It suggests why the jaguar, the puma, and forty-five species of birds have been extirpated from a place called Barro Colorado Island—and why myriad other creatures are mysteriously absent from myriad other sites. An ecosystem is a tapestry of species and relationships. Chop away a section, isolate that section, and there arises the problem of unraveling[87]."

Unlike drought, five million years ago the Congo Basin was green. In fact, it may have been wet - very wet. If we postulate that the separation of primate tribes occurred, not because of drought, but rather because of flood, we have a cause for both isolation and speciation.

Consider the proto-human waking up to a plateau once dry, now wet and marshy. If there was an ecological change, this would alter not only the activities and resources on the ground, but would affect the arboreal territory as well. Plant life would change. Large trees with shallow root systems would fail. The canopy that the early primates knew would dissipate. We would be forced to the ground. Suddenly we proto-humans would be bound to an island biogeography.

From a geographic standpoint it would require an alteration in habits. However, the primary resources for survival (note:water) would be present unlike that found in the drought scenario. With ample resources, we early humans would have to wade through the water - in an upright position - to get resources. Some of those resources would eventually be found along the waters edge.

Bonobos will wade into waters to feed on reeds and flowers. They have also been known to fish for shrimp with their hands. They crave water-lily flowers as a culinary treat and will wade deep when they spot one. Orangutans can swim and have been spotted swimming to migrate to trees in Indonesia. The proboscis monkey will swim a mile to go from one island to another. Snow monkeys will wade in water to take a bath and remove food particles from their fur.

"At least 10% of extant primates interact with aquatic environments, and a more complete understanding of these interactions is needed to get a complete view of primate behaviour. Five major factors appear to most strongly influence primate water use: thermoregulation, display behaviour, range, diet and predation[88]."

Far from the myth that primates do not like water, there are numerous examples of primate interaction with water: from creating drinking straws to using chewed leaves as sponges, primates are far from scared of the water.

One has to imagine the scenario playing out in an isolated are over an epoch of time and we have speciation and evolution. As the environment changed and the bipedal human was able to migrate, this new species never returned to its tree canopy, now a foreign land dominated by the "other" tree dwellers with whom the human now does not identify as "same." We became the "other." We became the new face in the primate parade. We became the water walkers. We became the upright monkeys with sore backs and corrective shoes. We became the osteoporosis outcasts who spends an inordinate amount for a small abode with a water view.

MESSAGE IN A BOTTLE

Research is what I'm doing when I don't know what I'm doing.
— Wernher von Braun

I had the benefit of being in the music gifted and talented program. On a field trip we went to the Kennedy Center when Leonard Bernstein was conducting an evening of Beethoven. They were rehearsing the ninth symphony and at one point Mr. Bernstein stopped the orchestra and he said, "Did you hear that?"

There was total silence.

"There is a countermelody that runs through the french horns." He turned to us. "Did you guys hear that?"

He reset the orchestra and at the appropriate time signaled the french horns who brought out this folk melody that otherwise would have been lost by the main melody of the strings.

"That's genius, right there."

The beauty of the composition was the intricacy of the melodies which intertwined to create a beautiful whole. With so many scientific theories cherry picking reasons for a specific biological modifications. But science, like a symphony, doesn't stand still. Its beauty is in its motion. It is not any one phrase that determines the work, it is its totality that renders the process pleasurable.

Science, itself, has been put on notice by a community who prefers not to wonder and postulate with a given set of facts. The ease with which many come to conclusions based on the opinions of others is tantamount to an inquisition which has far-reaching consequences for all of science.

The incorporation of science has prevented the free flow of ideas across fields and stagnated our ability to find insights and innovations that might help our society and the planet as a whole. Politicians, who are in charge of enacting policy are failing to consider the scientific community's data in its decisions.

In many ways, science has lost the very thing that has made it science. It is the random testing and accumulation of knowledge by observation and interactions, testing and manipulating, and for lack of a better word *play*.

I lived in a world where my golden five-speed bicycle with the long handle bars, banana seat and sissy bar would take through the landscape of Hood College in Frederick Maryland, Baker park in the heart of town, the Frederick library which smelled of old books and wisdom, and the no man's land of the East side of town by the airport.

It was a time where summers allowed me to go out and do what I wanted as long as I was back before dark. It is a different world now. There are more fences. There are more people. I'm not sure I'd thrive now if I was a child. Like those I see around me, I'd engage in the electronic virtual universe because that is the new social complex where I can be me or whatever avatar my creative mind might concoct.

Out there I learned about skunk cabbage. I learned that some ginkgo trees left droppings that smelled like feces. I set off model rockets. After setting off model rockets I make a boat propelled by model rockets and set it loose on the pond called Culler Lake much to the alarm of the neighborhood ducks that lived there. I played with sharp toys. I broke my arm, and still survived childhood. I never heard of food allergies other than a disdain for ketchup.

During the summer I would go to the Hood College pool and swim. I had a crush on a lifeguard named Muana. Her pixie hair had blond tips and she was tanned and always took time to talk to me. Of course I was only thirteen and she was an older woman of sixteen and a half with a drivers license. When not swimming I would read Ray Bradbury and Arthur Clarke, opening my mind to wonder of what might be. It was during the summer months where I felt I learned more than anytime in school. Even then, as I practiced swimming an entire lap in one breath, that we were meant for the water. I craved those times when we headed to the beach. Sometimes, depending on finance, it was a week long jaunt to the local shores. Other times it was a trip to Florida. When I arrived, I felt *home*.

Even then, I pretended I was part fish. Perhaps this has been a lifelong pursuit and this text is a realization of my quest. I am a writer, professional SCUBA diver, musician, and artist. I know the body in water. I know what it does.

When going for my deep certification as part of my Divemaster training, I submerged on the wreck the *Spiegel Grove*. It is a lovely wreck. Being the first one to the wheelhouse, I saw two behemoth Goliath Groupers eye me before disappearing into the blue void. Claude, my instructor took me over the side and we descended toward the sand. It was deeper than I've ever gone before. When you go that deep you get *Nitrogen Narcosis*. Some say it is the equivalent of the three martini lunch. It has been little understood. Some people lose their sense of inhibition. I've heard stories of people taking out their regulator, taking off their clothes or just wandering off and leaving their buddy.

"The effect of nitrogen on the body takes place in the central nervous system (CNS), but the exact site and mechanism are still debated. The lipid solubility hypothesis by Meyer and Overton noted that there is a correlation between the solubility of an anesthetic in lipid and its narcotic power. They also stated that "all gaseous and volatile substances induce narcosis if they penetrate cell lipids in a definite molar concentration which is characteristic for each type of cell." This theory was expanded by applying the "critical volume" concept which states that narcosis occurs when the inert gas or anesthetic changes a lipid portion of the cell. This is often thought to be the cell membrane, causing that portion of the cell to swell to a certain volume, impairing its function for that specific cell type[89]."

In fact, nitrogen is not the only inert gas that has an effect on the nervous system at depth. CO2 plays as important a role in the equation, multiplying the anesthetic effect.

"Carbon dioxide (CO2) is the gaseous end product of the aerobic metabolism of oxygen. CO2 is highly soluble in body tissues, and readily diffuses from cells to blood, where circulation transports it to the lungs for elimination. Divers often ignore carbon dioxide, as CO2 is a normal part of life. However, CO2 may have definite and detrimental effects if a diver accumulates an excessive amount of CO2…Carbon dioxide is a narcotic gas capable of depressing awareness to the degree of total loss of consciousness. In humans, acute elevation of arterial PCO2 above 70-75 mmHg reduces the level of awareness (20), and PaCO2 above 100-120 mmHg produces unresponsiveness (26). Severe elevation of PaCO2, by inhalation of 30%-40% CO2 (220-300 mmHg), produces surgical anesthesia in both animals and humans[90] (14,25)."

At the surface, prior to my descent on the *Spiegel Grove*, I was given a sheet with random squares numbered one to fifty. I was asked to point to each square in numerical order starting with one. While I scrambled to find the numbers in order, I was timed. At depth, after diving at depths that induce nitrogen narcosis, I was asked to perform the same operation, this time with a sheet containing the same number in a different random pattern. I was timed at depth. I felt sluggish, it was hard to concentrate.

Back at the surface they compared my times and found that I actually performed *better* at depth! How could this be? I explained that this was the "cop in the rear view mirror" syndrome. Perhaps it was my ability to write after a few glasses of rum. Like Hemingway said, "Write drunk, edit sober." Regardless, I was the anomaly in this test. I was the standard deviation, the point outside the curve. It is good to know that I was able to keep my wits at depth.

Much science is still out on apnea and SCUBA diving. We are still learning. I recently purchased a new dive computer. Its algorithm for decompression is far more conservative than anything else I owned. Needless to say, I've forgotten to take deep safety stops and wondered when "someone's" dive computer started chiming furiously.

It has become a ritual for my wife to get me to the ocean during the winter months. Even if it is a day to walk along the beach and collect a few shells, she knows that I get "corrected" by the smell of the sea air and the sound of the waves. During that time, I wonder what the protocol human thought when he first took a taste of a raw oyster, or dined on shrimp delicacies. I wonder if he used shells as tools. The beach provides me those moments to ponder a theory that, to me, makes perfect sense.

I can't help but wonder if we are, at some level, half-baked. When talking about evolution, one term that is often batted around is *convergent evolution*. Simply stated, given the same set of circumstances animals will evolve along similar paths. We have ants that look like beetles. Half a world away, we have lizards that are otherwise indistinguishable from each other. Whales and dolphins resemble fish. Noah's ark would have been filled with dozens of Dopplegangers, yet we stand alone in our uniqueness of bipedal critters. Could a lizard have developed to look like a human? It is possible. But we seem like the orphans of a misbegotten line of evolutionary experiments that still seems to hang on, even though we have no realm, no space, in which we are perfectly adapted

Or perhaps it would be more accurate to say we are half-evolved. I wonder what it would have been like had we not had to move from one territory to the next and give up our coastal evolutionary process. For those who may wonder, we are still evolving. We are still changing. Time and adaptation still takes place on humanity all over the world. In new and wondrous ways we are becoming a different creature than just a scant few thousand years ago.

The Coastal Theory scaffold of knowledge will be poked and prodded. In time, it may have to be revised as new information, specifically as genetics reveals more of who we are, comes to light.

What is evident is the internal reflection of the self as a primate unlike many other animals on earth. We have been changed by external and internal forces to become the vertical creature that we are now. We walk and swim and dance and sing because of something which differentiated us long ago.

We can conjecture to those times with the scant evidence laid before us. I propose the Coastal Theory because I exist in a state that supports this theory. I can stand on my toes. I am a hairless sweaty beast who communicates verbally and hunts visually. I can hold my breath while submerging to depths many thought dangerous only a few years ago.

If you find yourself along the beach as you pick up shells and examine them for their intricacy and beauty, stop and ponder how many have done so before you. You are not the only one on the beach. For millions of years there were others, not so dissimilar from you who may have felt the soft wet sand between their toes. They waded into the cool water and back out again with ease. They found a moment to stop between the two worlds and looked up and marveled at the lights in the sky as the rhythm of the ocean waves crashed along the shore reminding us that the engine of Earth continues to move unceasingly forward. It is the clockwork of life that moves at its own pace, and calls us to do the same.

ABOUT THE AUTHOR

It had diminished him to nothingness. Now he could feel. As if awakened from a deep sleep he was alive and refreshed. He could live again. He could love with a depth he had thought had been lost forever.

- M. T. Harber - Sweet Taste of the Bilge

M.T. Harber (a.k.a Paul Rose) is a curious soul who enjoys traveling whenever he can afford it. Obsessed with the watery world, M.T. is currently a PADI divemaster and is training to be an Open Water SCUBA Instructor for SSI.

Born in Frederick Maryland he used to get kicked out of public pools when he practiced his breath holding techniques to the chagrin of life-guards who'd thought he'd drowned. He's sailed the Chesapeake, dove wrecks in the Florida Keys, explored reefs in the Caribbean Sea, and intertwined maritime research into vivid novels rich with salty air.

Besides writing, M.T. Enjoys painting and drawing. His commissioned works are sought after in the dive community. In particular, his paintings of turtles have garnered much praise. He's an award winning musician, having won three Telly awards for music he composed for television.

BIBLIOGRAPHY

Olena, A, 'Human-Specific Genes Implicated in Brain Size', The Scientist 2018,

McVean, A, 'Whiskers on Humans', Office for Science and Society McGill

adolescent ..., SFB-NDFCA, & 2011, 'Property in nonhuman primates', Wiley Online Library

Akitomo, Y, H Akamatsu, Y Okano, HM-JO dermatological ..., & 2003, 'Effects of UV irradiation on the sebaceous gland and sebum secretion in hamsters', Elsevier

Alexander, RM, 'Bipedal animals, and their differences from humans.', J Anat vol. 204, no. 5, 2004, pp. 321-330.

Bastir, M, A Rosas, P Gunz, AP-M Nature ..., & 2011, 'Evolution of the base of the brain in highly encephalized human species', nature.com

Becker, BE, KH-...O Aquatic ..., & 2009, 'Biophysiologic effects of warm water immersion', scholarworks.bgsu.edu

Biniek, K, K Levi, & RH Dauskardt, 'Solar UV radiation reduces the barrier function of human skin.', Proc Natl Acad Sci U S A vol. 109, no. 42, 2012, pp. 17111-17116.

Bovell, DL, 'The evolution of eccrine sweat gland research towards developing a model for human sweat gland function.', Exp Dermatol vol. 27, no. 5, 2018, pp. 544-550.

Breslin, PA, 'An evolutionary perspective on food and human taste.', Curr Biol vol. 23, no. 9, 2013, pp. R409-18.

Charles, HS, & B George, 'Natural Selection and Beyond', 2010,

Charles, M, 'The Shark God', 2009,

Barras, C, 'The evolution of the nose: why is the human hooter so big?', New Scientist 2016,

com, DH-S, December, & 2016, 'How the Story of "Moana" and Maui Holds Up Against Cultural Truth', Smithsonian National Museum of Natural History

Darwin, C, The Origin of Species, CreateSpace Independent Publishing Platform, 2018.

Abbott, D, 'What brain regions control our language? And how do we know this?', The Conversation vol. Sept, 2016, 2016,

Dávid-Barrett, T, & RI Dunbar, 'Bipedality and hair loss in human evolution revisited: The impact of altitude and activity scheduling.', J Hum Evol vol. 94, 2016, pp. 72-82.

David, Q, 'The Song Of The Dodo', 2012,

Diogo, R, JL Molnar, & B Wood, 'Bonobo anatomy reveals stasis and mosaicism in chimpanzee evolution, and supports bonobos as the most appropriate extant model for the common ancestor of chimpanzees and humans.', Sci Rep vol. 7, no. 1, 2017, pp. 608.

Ecott, T, Neutral Buoyancy: Adventures in a Liquid World, Grove Press, 2002.

education, LAP-N, & 2008, 'Transposons: The jumping genes', molgen.biologie.uni-mainz.de

'Fact Sheet: The Fragrance Industry's Policy Failures and Trade Secret Myth', Women's Voices for the Earch

Gee, H, The Accidental Species: Misunderstandings of Human Evolution, Reprint edn., University Of Chicago Press, 2015.

GerardD.Gierlinskia, GrzegorzNiedzwiedzkib, MartinG.Lockley, AthanassiosAthanassioue, CharalamposFassoulasf, ZofiaDubickag,AndrzejBoczarowskic, MatthewR.Bennettk, & PerErikAhlberg,

'Possible hominin footprints from the late Miocene(c.5.7Ma) of Crete?', Proceedings of theGeologists' Association vol. 128, no. 2017, 2017, pp. 697-710.

Germonpré, P, C Balestra, PM-BJO sports ..., & 2011, 'Passive flooding of paranasal sinuses and middle ears as a method of equalisation in extreme breath-hold diving', bjsm.bmj.com

Gilsanz, V, HH Hu, & S Kajimura, 'Relevance of brown adipose tissue in infancy and adolescence.', Pediatr Res vol. 73, no. 1, 2013, pp. 3-9.

Grehan, JR, JHS-JO Biogeography, & 2009, 'Evolution of the second orangutan: phylogeny and biogeography of hominid origins', Wiley Online Library

Hatala, KG, B Demes, & BG Richmond, 'Laetoli footprints reveal bipedal gait biomechanics different from those of modern humans and chimpanzees.', Proc Biol Sci vol. 283, no. 1836, 2016,

Horrocks, LA, YKY-P research, & 1999, 'Health benefits of docosahexaenoic acid (DHA)', Elsevier

Hutchinson, S, 'What's the Difference Between Hair and Fur?', Mental Floss 2014,

Ilardo, MA, I Moltke, TS Korneliussen, J Cheng, AJ Stern, F Racimo, P de Barros Damgaard, M Sikora, A Seguin-Orlando, S Rasmussen, ICL van den Munckhof, R Ter Horst, LAB Joosten, MG Netea, S Salingkat, R Nielsen, & E Willerslev, 'Physiological and Genetic Adaptations to Diving in Sea Nomads.', Cell vol. 173, no. 3, 2018, pp. 569-580.e15.

Jablonski, NG, 'The naked truth.', Sci Am vol. 302, no. 2, 2010, pp. 42-49.

Jablonski, NG, GC-D clinics, & 2014, 'The evolution of skin pigmentation and hair texture in people of African ancestry', jaad.org

Jokiel, PL, KS Rodgers, WJW-JO Marine ..., & 2011, 'Marine resource management in the Hawaiian Archipelago: the traditional Hawaiian system in relation to the Western approach', hindawi.com

K. D'Août, KL, B. Van Gheluwe & D. De Clercq, Advances in Plantar Pressure Measurements in Clinical and Scientific Research, Shaker Publishing Shaker Publishing, BV,

Kang, KS, JM Hong, DJ Horan, KE Lim, WA Bullock, A Bruzzaniti, S Hann, ML Warman, & AG Robling, 'Induction of Lrp5 HBM-causing mutations in Cathepsin-K expressing cells alters bone metabolism.', Bone vol. 120, 2018, pp. 166-175.

Kean, S, The Violinist's Thumb: And Other Lost Tales of Love, War, and Genius, as Written by Our Genetic Code, Reprint edn., Back Bay Books, 2013.

King, GE, Primate Behavior and Human Origins, 1 edn., Routledge, 2015.

Kiyatkin, EA, 'Brain temperature homeostasis: physiological fluctuations and pathological shifts.', Front Biosci (Landmark Ed) vol. 15, 2010, pp. 73-92.

Lee, S-H, Close Encounters with Humankind: A Paleoanthropologist Investigates Our Evolving Species, 1 edn., W. W. Norton & Company, 2018.

Masuda, K, K Okazaki, S Kuno, K Asano, H Shimojo, & S Katsuta, 'Endurance training under 2500-m hypoxia does not increase myoglobin content in human skeletal muscle.', Eur J Appl Physiol vol. 85, no. 5, 2001, pp. 486-490.

McElwee, KJ, D Boggess, JM-JO Investigative ..., & 1999, 'Spontaneous alopecia areata-like hair loss in one congenic and seven inbred laboratory mouse strains', core.ac.uk

Mojumdar, EH, QD Pham, D Topgaard, & E Sparr, 'Skin hydration: interplay between molecular dynamics, structure and water uptake in the stratum corneum.', Sci Rep vol. 7, no. 1, 2017, pp. 15712.

Morgan, E, Aquatic Ape Hypothesis, New edition edn., Souvenir Press, 1982.

Morgan, E, The Aquatic Ape, First Edition edn., Stein & Day Pub, 1982.

Morgan, E, The Descent of Woman: The Classic Study of Evolution, Fourth Edition, Fourth edition edn., Souvenir Press, 2001.

Morris, D, The Naked Ape: A Zoologist's Study of the Human Animal, Delta, 1999.

Anderson, N, Chimpanzees, Orangutans Can Swim and Dive, Biologists Prove, 2013, retrieved 7/3/2016 2016, <http://www.sci-news.com/biology/science-chimpanzees-orangutans-swim-dive-01319.html>.

Administration, NOAA, What percentage of the American population lives near the coast?, <https://oceanservice.noaa.gov/facts/population.html>.

Nestor, J, Deep: Freediving, Renegade Science, and What the Ocean Tells Us About Ourselves, Reprint edn., Eamon Dolan/Mariner Books, 2015.

Staff, N, '5.6-Million-Year-Old Hominin-Like Footprints in Crete Challenge Theories of Human Evolution', Science News 2017,

Nichols, WJ, Blue Mind: The Surprising Science That Shows How Being Near, In, On, or Under Water Can Make You Happier, Healthier, More Connected, and Better at What You Do, Reprint edn., Back Bay Books, 2015.

Davis, N, 'David Attenborough's aquatic ape series for Radio 4 based on 'wishful thinking'', The Guardian 2016,

Norsk, P, FB-P-JO Applied ..., & 1990, 'Catecholamines, circulation, and the kidney during water immersion in humans', physiology.org

O'Connor, RE, 'Water intoxication with seizures.', Ann Emerg Med vol. 14, no. 1, 1985, pp. 71-73.

Qin, X, HG Park, JY Zhang, P Lawrence, G Liu, N Subramanian, KS Kothapalli, & JT Brenna, 'Brown but not white adipose cells synthesize omega-3 docosahexaenoic acid in culture.', Prostaglandins Leukot Essent Fatty Acids vol. 104, 2016, pp. 19-24.

Rae, TC, TK-EA Issues, News ..., & 2014, 'Sinuses and flotation: Does the aquatic ape theory hold water', Wiley Online Library

Khan, R, 'Humans as the aquaphilic Ape', Discover Magazine 2013,

Review, WTF-PB, & 2017, 'Empirical approaches to the study of language evolution', Springer

Sacks, O, Island of the Colour-blind, Picador, 1997.

Semaw, S, SW Simpson, J Quade, PR Renne, RFB Nature, & 2005, 'Early Pliocene hominids from Gona, Ethiopia', nature.com

Shubin, N, Your Inner Fish: A Journey into the 3.5-Billion-Year History of the Human Body, 1 Reprint edn., Vintage, 2009.

Stephen, C, & S Kathlyn, 'Human Brain Evolution', 2010,

Folger, T, 'The Naked and the Bipedal', Discover Magazine 1993,

Vahter, M, G Concha, B Nermell, RN-EJ of ..., & 1995, 'A unique metabolism of inorganic arsenic in native Andean women', Elsevier

WADE, NICHOLAS, 'Why Humans and Their Fur Parted Ways ', The New York Times 2003, pp. 5.

What does it mean to be Human?, retrieved May 28, 2018 2018, <http://humanorigins.si.edu/evidence/human-fossils/species/ardipithecus-kadabba>.

Wilson, CJ, M Das, S Jayaraman, O Gursky, & JR Engen, 'Effects of Disease-Causing Mutations on the Conformation of Human Apolipoprotein A-I in Model Lipoproteins.', Biochemistry vol. 57, no. 30, 2018, pp. 4583-4596.

Wrangham, R, D Cheney, R Seyfarth, & E Sarmiento, 'Shallow-water habitats as sources of fallback foods for hominins.', Am J Phys Anthropol vol. 140, no. 4, 2009, pp. 630-642.

Zihlman, AL, DRB-POT National ..., & 2015, 'Body composition in Pan paniscus compared with Homo sapiens has implications for changes during human evolution', National Acad Sciences

Zoology, MJR-JO, & 2007, 'Evolution of nakedness in Homo sapiens, Wiley Online Library

FOOTNOTES

1 (Morris, D, The Naked Ape: A Zoologist's Study of the Human Animal, Delta, 1999.)

2 (Morgan, E, The Descent of Woman: The Classic Study of Evolution, Fourth Edition, Fourth edition edn., Souvenir Press, 2001.)

3 (Ibid.)

4 (Morgan, E, The Aquatic Ape, First Edition edn., Stein & Day Pub, 1982.) p31

5 (Gee, H, The Accidental Species: Misunderstandings of Human Evolution, Reprint edn., University Of Chicago Press, 2015.)

6 (NG Jablonski, 'The naked truth.', Sci Am vol. 302, no. 2, 2010, pp. 42-49.)

7 (K. D'Août, KL, B. Van Gheluwe & D. De Clercq, Advances in Plantar Pressure Measurements in Clinical and Scientific Research, Shaker Publishing Shaker Publishing, BV,)

8 The Miocene "Proconsul" is commonly viewed as a precursor to the ape, chimpanzee split. Note the name Proconsul came from a popular circus chimp famous for riding a bicycle and smoking cigarettes.

9 (What does it mean to be Human?, retrieved May 28, 2018 2018, <http://humanorigins.si.edu/evidence/human-fossils/species/ardipithecus-kadabba>.)

10 (KG Hatala, B Demes, & BG Richmond, 'Laetoli footprints reveal bipedal gait biomechanics different from those of modern humans and chimpanzees.', Proc Biol Sci vol. 283, no. 1836, 2016,)

11 (N Staff, '5.6-Million-Year-Old Hominin-Like Footprints in Crete Challenge Theories of Human Evolution', Science News 2017,)

12 (Lee, S-H, Close Encounters with Humankind: A Paleoanthropologist Investigates Our Evolving Species, 1 edn., W. W. Norton & Company, 2018.)

13 (RM Alexander, 'Bipedal animals, and their differences from humans.', J Anat vol. 204, no. 5, 2004, pp. 321-330.)

14 (King, GE, Primate Behavior and Human Origins, 1 edn., Routledge, 2015.)

15 (T Folger, 'The Naked and the Bipedal', Discover Magazine 1993,)

16 (T Dávid-Barrett, & RI Dunbar, 'Bipedality and hair loss in human evolution revisited: The impact of altitude and activity scheduling.', J Hum Evol vol. 94, 2016, pp. 72-82.)

17 (EA Kiyatkin, 'Brain temperature homeostasis: physiological fluctuations and pathological shifts.', Front Biosci (Landmark Ed) vol. 15, 2010, pp. 73-92.)

18 (N Davis, 'David Attenborough's aquatic ape series for Radio 4 based on 'wishful thinking'', The Guardian 2016,)\

19 (H Tuomisto, M Tuomisto, & JT Tuomisto, 'How scientists perceive the evolutionary origin of human traits: Results of a survey study.', Ecol Evol vol. 8, no. 6, 2018, pp. 3518-3533.)

20 (S Semaw, SW Simpson, J Quade, PR Renne, RFB Nature, & 2005, 'Early Pliocene hominids from Gona, Ethiopia', nature.com)

21 (GerardD.Gierlinskia, GrzegorzNiedzwiedzkib, MartinG.Lockley, AthanassiosAthanassioue, CharalamposFassoulasf, ZofiaDubickag,AndrzejBoczarowskic, MatthewR.Bennettk, & PerErikAhlberg, 'Possible hominin footprints from the late Miocene(c.5.7Ma) of Crete?', Proceedings of theGeologists' Association vol. 128, no. 2017, 2017, pp. 697-710.)

22 (Anderson, N, Chimpanzees, Orangutans Can Swim and Dive, Biologists Prove, 2013, retrieved 7/3/2016 2016, <http://www.sci-news.com/biology/science-chimpanzees-orangutans-swim-dive-01319.html>.)

23 An interesting point with the sea nomads, specifically the Baja Laut, is that they retain an almost exclusive aquatic life. Their homes are boats, and their

source of income is fishing. Those born on their boat-homes retain no citizen ship to any country which poses issues when they have to come ashore. Add to that the increased dynamite fishing which is devastating their livelihood.

24 (MA Ilardo, I Moltke, TS Korneliussen, J Cheng, AJ Stern, F Racimo, P de Barros Damgaard, M Sikora, A Seguin-Orlando, S Rasmussen, ICL van den Munckhof, R Ter Horst, LAB Joosten, MG Netea, S Salingkat, R Nielsen, & E Willerslev, 'Physiological and Genetic Adaptations to Diving in Sea Nomads.', Cell vol. 173, no. 3, 2018, pp. 569-580.e15.)

25 (Ecott, T, Neutral Buoyancy: Adventures in a Liquid World, Grove Press, 2002.) Pg 8

26 (Nestor, J, Deep: Freediving, Renegade Science, and What the Ocean Tells Us About Ourselves, Reprint edn., Eamon Dolan/Mariner Books, 2015.) Pg 7

27 You want to know how to cure hiccups? Think of the epiglottis as a muscle that is cramping. The best way to cure a cramp is to massage it. No, I'm not telling you to stick your finger down your throat. However, a tablespoon of sugar (you could use any grainy alternative - sand for example) swallowed will roll down the epiglottis and is usually enough to calm its nervous "twitch."

Also, this trait apparently comes from a period where we had much in common with tadpoles - who also have hiccups to help them breathe!

28 (C Barras, 'The evolution of the nose: why is the human hooter so big?', New Scientist 2016,)

29 (Anderson, Chimpanzees, Orangutans Can Swim and Dive, Biologists Prove.)

30 (K Masuda, K Okazaki, S Kuno, K Asano, H Shimojo, & S Katsuta, 'Endurance training under 2500-m hypoxia does not increase myoglobin content in human skeletal muscle.', Eur J Appl Physiol vol. 85, no. 5, 2001, pp. 486-490.)

31 (Nestor, Deep: Freediving, Renegade Science, and What the Ocean Tells Us About Ourselves.)

32 If you have a newborn, you can try this experiment with them. Blow across their face and they will gasp for air as well. The mammalian reflex is alive and well in them.

33 (Administration, NOAA, What percentage of the American population lives near the coast?, <https://oceanservice.noaa.gov/facts/population.html>.)

34 (Darwin, C, The Origin of Species, CreateSpace Independent Publishing Platform, 2018.)

35 (NICHOLAS WADE, 'Why Humans and Their Fur Parted Ways ', The New York Times 2003, pp. 5.)

36 (Dávid-Barrett, & Dunbar, Bipedality and hair loss in human evolution revisited: The impact of altitude and activity scheduling.)

37 (MJR-JO Zoology, & 2007, 'Evolution of nakedness in Homo sapiens', Wiley Online Library)

38 (A McVean, 'Whiskers on Humans', Office for Science and Society McGill)

39 (S Hutchinson, 'What's the Difference Between Hair and Fur?', Mental Floss 2014,)

40 (NG Jablonski, GC-D clinics, & 2014, 'The evolution of skin pigmentation and hair texture in people of African ancestry', jaad.org)

41 (Jablonski, The naked truth.)

42 ('Fact Sheet: The Fragrance Industry's Policy Failures and Trade Secret Myth', Women's Voices for the Earch)

43 (DL Bovell, 'The evolution of eccrine sweat gland research towards developing a model for human sweat gland function.', Exp Dermatol vol. 27, no. 5, 2018, pp. 544-550.)

44 (Morgan, E, Aquatic Ape Hypothesis, New edition edn., Souvenir Press, 1982.)

45 (AL Zihlman, DRB-POT National ..., & 2015, 'Body composition in Pan paniscus compared with Homo sapiens has implications for changes during human evolution', National Acad Sciences)

46 (Y Akitomo, H Akamatsu, Y Okano, HM-JO dermatological ..., & 2003, 'Effects of UV irradiation on the sebaceous gland and sebum secretion in hamsters', Elsevier)

47 (RE O'Connor, 'Water intoxication with seizures.', Ann Emerg Med vol. 14, no. 1, 1985, pp. 71-73.)

48 (X Qin, HG Park, JY Zhang, P Lawrence, G Liu, N Subramanian, KS Kothapalli, & JT Brenna, 'Brown but not white adipose cells synthesize omega-3 docosahexaenoic acid in culture.', Prostaglandins Leukot Essent Fatty Acids vol. 104, 2016, pp. 19-24.)

49 (V Gilsanz, HH Hu, & S Kajimura, 'Relevance of brown adipose tissue in infancy and adolescence.', Pediatr Res vol. 73, no. 1, 2013, pp. 3-9.)

50 (EH Mojumdar, QD Pham, D Topgaard, & E Sparr, 'Skin hydration: interplay between molecular dynamics, structure and water uptake in the stratum corneum.', Sci Rep vol. 7, no. 1, 2017, pp. 15712.)

(K Biniek, K Levi, & RH Dauskardt, 'Solar UV radiation reduces the barrier function of human skin.', Proc Natl Acad Sci U S A vol. 109, no. 42, 2012, pp. 17111-17116.)

51 (JR Grehan, JHS-JO Biogeography, & 2009, 'Evolution of the second orangutan: phylogeny and biogeography of hominid origins', Wiley Online Library)

52 (PA Breslin, 'An evolutionary perspective on food and human taste.', Curr Biol vol. 23, no. 9, 2013, pp. R409-18.)

53 (M Bastir, A Rosas, P Gunz, AP-M Nature ..., & 2011, 'Evolution of the base of the brain in highly encephalized human species', nature.com)

54 It is true that the nose contains magnetite (metals that orient to the poles) and is used as a means for sensing direction. It is a built in compass within the human anatomy. Sadly, for those like me, I must be totally bereft of this substance having set out in one direction and ending up in the wrong state!

55 (Shubin, N, Your Inner Fish: A Journey into the 3.5-Billion-Year History of the Human Body, 1 Reprint edn., Vintage, 2009.)

56 (TC Rae, TK-EA Issues, News ..., & 2014, 'Sinuses and flotation: Does the aquatic ape theory hold water', Wiley Online Library)

57 Failure to purify the water introduces a host of microbes to the liquid areas of the sinus and the brain. Brain-eating amoebas can lodge and grow without properly processing the fluid.

58 (P Germonpré, C Balestra, PM-BJO sports ..., & 2011, 'Passive flooding of paranasal sinuses and middle ears as a method of equalisation in extreme breath-hold diving', bjsm.bmj.com)

59 (WTF-PB Review, & 2017, 'Empirical approaches to the study of language evolution', Springer)

60 (D Abbott, 'What brain regions control our language? And how do we know this?', The Conversation vol. Sept, 2016, 2016,)

61 (Shubin, Your Inner Fish: A Journey into the 3.5-Billion-Year History of the Human Body.), p 90.

62 (R Diogo, JL Molnar, & B Wood, 'Bonobo anatomy reveals stasis and mosaicism in chimpanzee evolution, and supports bonobos as the most appropriate extant model for the common ancestor of chimpanzees and humans.', Sci Rep vol. 7, no. 1, 2017, pp. 608.)

63 Care needs to be taken when calculating these numbers. Some scientists argue that certain duplicated groups of DNA have been omitted and missing sequences not factored into the calculation. Therefore, the greatest drift could actually be calculated at about 4%. Nonetheless, this is still an incredible high correlation across the genetic spectrum.

64 (R Wrangham, D Cheney, R Seyfarth, & E Sarmiento, 'Shallow-water habitats as sources of fallback foods for hominins.', Am J Phys Anthropol vol. 140, no. 4, 2009, pp. 630-642.)

65 (KS Kang, JM Hong, DJ Horan, KE Lim, WA Bullock, A Bruzzaniti, S Hann, ML Warman, & AG Robling, 'Induction of Lrp5 HBM-causing mutations in Cathepsin-K expressing cells alters bone metabolism.', Bone vol. 120, 2018, pp. 166-175.)

66 (CJ Wilson, M Das, S Jayaraman, O Gursky, & JR Engen, 'Effects of Disease-Causing Mutations on the Conformation of Human Apolipoprotein A-I in Model Lipoproteins.', Biochemistry vol. 57, no. 30, 2018, pp. 4583-4596.)

67 (Sacks, O, Island of the Colour-blind, Picador, 1997.)

68 (M Vahter, G Concha, B Nermell, RN-EJ of ..., & 1995, 'A unique metabolism of inorganic arsenic in native Andean women', Elsevier)

69 (LAP-N education, & 2008, 'Transposons: The jumping genes', molgen.biologie.uni-mainz.de)

70 (Kean, S, The Violinist's Thumb: And Other Lost Tales of Love, War, and Genius, as Written by Our Genetic Code, Reprint edn., Back Bay Books, 2013.)

71 (KJ McElwee, D Boggess, JM-JO Investigative ..., & 1999, 'Spontaneous alopecia areata-like hair loss in one congenic and seven inbred laboratory mouse strains', core.ac.uk)

72 (A Olena, 'Human-Specific Genes Implicated in Brain Size', The Scientist 2018,)

73 (LA Horrocks, YKY-P research, & 1999, 'Health benefits of docosahexaenoic acid (DHA)', Elsevier)

74 (C Stephen, & S Kathlyn, 'Human Brain Evolution', 2010,)

75 (Ibid.)

76 (SFB-NDFCA adolescent ..., & 2011, 'Property in nonhuman primates', Wiley Online Library)

77 (M Charles, 'The Shark God', 2009,)

78 (DH-S com, December, & 2016, 'How the Story of "Moana" and Maui Holds Up Against Cultural Truth', Smithsonian National Museum of Natural History)

79 (PL Jokiel, KS Rodgers, WJW-JO Marine ..., & 2011, 'Marine resource management in the Hawaiian Archipelago: the traditional Hawaiian system in relation to the Western approach', hindawi.com)

80 (Nichols, WJ, Blue Mind: The Surprising Science That Shows How Being Near, In, On, or Under Water Can Make You Happier, Healthier, More Connected, and Better at What You Do, Reprint edn., Back Bay Books, 2015.)

81 (P Norsk, FB-P-JO Applied ..., & 1990, 'Catecholamines, circulation, and the kidney during water immersion in humans', physiology.org)

82 (BE Becker, KH-...O Aquatic ..., & 2009, 'Biophysiologic effects of warm water immersion', scholarworks.bgsu.edu)

83 (R Khan, 'Humans as the aquaphilic Ape', Discover Magazine 2013,)

84 (BL Jonathan, 'Improbable Destinies', 2017,)

85 (Q David, 'The Song Of The Dodo', 2012,)

86 (HS Charles, & B George, 'Natural Selection and Beyond', 2010,)

87 (David, The Song Of The Dodo.)

88 (E Kempf, 'Patterns of water use in primates.', Folia Primatol (Basel) vol. 80, no. 4, 2009, pp. 275-294.)

89 (PJ Kirkland, & JS Cooper, 'Nitrogen Narcosis In Diving' in (eds.), StatPearls, StatPearls Publishing, Treasure Island (FL), 2019,)

90 (JE Brian, Jr MD, 'Carbon Diaoxide, Narcosis, and Diving', Global Underwater Explorers)

Made in the USA
Columbia, SC
24 July 2019